HOME LIGHTING

Also by Anthony Byers

HOME ELECTRICITY

HOME LIGHTING

Anthony Byers

PELHAM BOOKS

First published in Great Britain by
PELHAM BOOKS LTD
52 *Bedford Square*
*London, W.C.*1
November 1970
First Paperback Edition 1971

7207 0382 4

Set and printed in Great Britain by
Tonbridge Printers Ltd, Peach Hall Works, Tonbridge, Kent
in Baskerville eleven on thirteen point on paper supplied by
P. F. Bingham Ltd, and bound by James Burn
at Esher, Surrey

All figures and measurements in this book are in metric and SI units

METRIC CONVERSION TABLE

Metric units have been adopted in line with their acceptance by the makers of lamps and fittings and the electrical industry. This table is for the conversion of figures used in these pages.

Length

1 millimetre (mm)	=	0.03937 in.
1 inch	=	25.4 mm
1 metre (m)	=	39.37 inches or 3.28 ft.
1 foot	=	0.3048 m
1 yard	=	0.9144 m

Area

1 square metre (m^2)	=	10.764 sq. ft.
1 square foot	=	0.0920 m^2
1 square yard	=	0.836 m^2

Weight

1 gramme (g)	=	0.0352 ounce.
1 ounce	=	28.35 g.
1 kilogramme (Kg)	=	2.2046 lb.
1 pound (lb) (K)	=	0.453 kg

Illumination

1 Lux = 0.093 lumens per sq. ft $1m/ft^2$ or Foot-candle.

1 lumen per sq. ft (or foot candle) = 10.76 Lux.

CONTENTS

ILLUSTRATIONS

Grateful acknowledgement is given to colleagues in The Electricity Council for their help – in particular to Mr John Watkinson, M.A.S.E.E., A.M.I.T.E., Technical Officer in the Engineering Services Section, and Mr Derek Durrant, F.Illum.E.S., Senior Lighting Engineer, whose home – Gore Hill – features among the illustrations of good lighting. The assistance of the leading makers of lamps and light fittings – and many of the younger, small and struggling firms with good ideas – is also recognised. But for the progressive marketing policies of these firms and their concern with good design this book would not be possible. Thanks are also due to J. A. Crabtree & Co. Ltd., for permission to use drawings from their electrical handbook. Most of all I am grateful to my wife who did the really hard work.

Chapter 1: LETTING IN THE LIGHT

Lighting should be far more than just a means of seeing at night. A room can be transformed by good lighting; it can add immeasurably to the pleasure we take in our homes, in our furniture, decorations, ornaments, glass and china. Yet most of us start with a home that is provided with the bare minimum of lighting points: one in the centre of a ceiling and a socket outlet in a wall which, in a living room, is expected to serve the TV set as well as a table or standard lamp.

Acceptance of a low standard of domestic lighting has been put down to a Puritan conscience. Virtue resides with those who are early to bed and early to rise; wickedness is associated with darkness, 'night-life' is synonymous with sin: though the imaginative scene of a den of vice is always poorly lit by a single shaded lamp or just a candle set crookedly in a bottle. A more likely explanation is the erroneous belief that lighting is expensive to buy and expensive to use, linked with fears about electricity which – because it cannot be seen – seems mysterious and difficult to understand.

Lighting is an essential part of decoration – as important as the wall or floor coverings. For a good part of each day we must depend on electric lighting, so it does seem reasonable to make the most of the opportunities open to us. Light fittings are probably the least expensive of any decoration for the home and there is now a great variety for a great many uses. Light can be linked with music so the sounds create colourful visual images; light can be used to bring out the beauty of a painting or a work of sculpture; light can transform a suburban garden into a wonderland of colour. There is no special mystique. It does not require any special technical ability to understand the decorative possibilities. It will not even matter if you fail to understand the terms of measurement or any of the technical jargon of the illuminating engineer. It will, though, help if you can understand some of the basic principles of home wiring since proper wiring is esential for good lighting. Also, a knowledge of home electricity will enable you to make convenient use of the many electric labour saving appliances for the home.

ELECTRICAL TERMS

So as to understand the essentials: electric current flows along the cables in the home at a fixed rate. The *rate* of flow is called the AMPERAGE. The *pressure* of flow is called the VOLTAGE. The rate at which a lamp or a piece of electrical equipment uses electricity is called the WATTAGE. If you think of a water pipe you will get the general idea. The weight of water in the tank in your roof exerts a certain amount of force or pressure (usually about 8 lb per square inch). This is equivalent to the voltage. But then the size of the pipe makes a difference. If the pressure is the same, you might expect a 1-inch pipe to fill a bath twice as quickly as a ½-inch pipe. But if the pressure were increased then the ½-inch pipe could be assumed to do it just as quickly. So, adapting our simile to electricity, you might say that the size of the conducting wire is related to the amperage.

Then there is the tap at the end of the pipe. If it is turned full on it will pass more water in a given time than if it is half turned on. So, again, you could imagine that the amount the tap is turned on is roughly equivalent to the wattage of an electrical appliance.

The *voltage* in most homes in Great Britain is 240 volts.

Some houses have a 200 or 210 or 220 volts supply but everyone will eventually have a standard 240 volts supply.

The other common voltage used is 110, found in many countries abroad. It has the advantage of being safer but the disadvantage of needing thicker cables in order to get the higher amperages needed by appliances with heating elements. It is, however, suitable for lighting which has a low amperage and in situations where special care must be taken – as with garden lighting – 110 volts is often used. Low voltages such as 12 and 24V are also used for some types of lamp.

The number of *amps* taken will depend upon the resistance (*ohms* is the term the electricians like to use) of the lamp or the appliance. If you multiply the amps by the volts you will be able to tell the amount of power the appliance or the lamp uses. Almost all electrical appliances give this information on a small plate or label fixed often to the back, or underneath, or behind the connection plate. If you want to work it out: multiply amps by volts and the answer is in watts (W). Or divide the volts into the watts and the answer is in number of amps. Fuse wire, circuit breakers, sockets and plugs are always measured in amps. *Lamps*, of course, are always measured in watts. The wattage printed on the glass or on the cap and on most light fittings specifies the wattage of the lamp that can be safely used in the holder. *Bulbs* for torches are measured in volts. They use so little electricity by comparison with most other lamps that it is usually easier to rate them in this way rather than in fractions of an amp.

The word 'bulb' is commonly used to refer to all incandescent light sources but, strictly speaking, this is a description of its shape and these days, with so many different shapes, it is more correct to call them lamps and limit 'bulbs' to the battery lights. In this book we will talk about 'lamps' but if you want to call them bulbs don't let us stop you.

The shopkeeper and the electrician will know what you are talking about.

COST OF LIGHTING

Because it uses so little electricity lighting is inexpensive; 1p. of electricity will light a 100 watt lamp for 13 hours, an 80W fluorescent lamp for 13 hours. The average electricity bill for lighting a three-bedroomed house is about £3 a *year*. For about £4 a year running costs the most wonderful improvements can be made.

The average cost of running an appliance on the normal domestic tariff in England and Wales is about 0.75 new pence *per unit*, depending on the prices fixed by the Electricity Board in the region. You can find out the exact price by looking at your electricity bill under the heading 'price per unit' or you can ask at an Electricity Board shop.

The standard *unit* of electrical consumption is the *kilowatt* hour (kWh) which is 1,000 watts used continuously for one hour. So ten lamps, each of 100 watts, burning for one hour would use one unit of electricity at an average price of 0.75p.

WIRING FOR LIGHTING

Electricity comes into the home through a service cable which goes into a sealed fuse termination and then to the electricity meter, then to the consumer unit which is commonly a square box containing fuses for the lighting circuits and for the socket outlets. In some homes the lighting circuit has a separate 'consumer unit' with its own switch and fuse.

In the early days of electricity one price was charged for the electricity used for lighting and another charge was made for power. So that householders could not use the lighting circuit for appliances like electric fires, different plugs and sockets and circuits were designed and this is one reason for the confusing number of different types of fuse boxes, plugs and sockets found in many older homes.

The modern way is to have a consumer unit with a safety on/off switch and a number of separate circuits, each controlled by a fuse. The fuse is the deliberately designed weak link in the circuit that breaks or melts if the circuit current increases above the amount the wiring is designed to carry.

Each fuse supplies one circuit exclusively. No other circuit should be supplied from any one fuse. The fuses, and therefore the circuits they supply, are made in four standard ratings: 5 amp, 15 amp, 20 amp, and 30 amp. Some consumer units include a 45 amp fuse for supplying a cooker circuit. A switch and a 60 amp circuit fuse can be supplied for special heavy duty circuits.

The current rating of the *main switch* ranges from 30 amps for the smallest unit, to 60 amps or 80 amps. The ratings vary with the make but with the increasing load demand of many homes using low cost electricity for heating and water heating some consumer units are now being made with a main switch of 80 amp or, occasionally, 100 amp rating controlled by a separate switch.

Also near the consumers unit is an earth terminal to which all the circuit earth conductors in the house must be connected. One wire in the three core electric cable is the earth conductor. It is usually green (or green and yellow) sheathed and sometimes it is bare tinned copper wire. Until 1966 lighting circuits did not have an earth wire but now it is considered essential in the interests of safety to provide an earth lead connection, for a great many modern lighting fittings are made of metal which is a good conductor of electricity. If any metal part of the lamp were accidentally to come into contact with an uninsulated portion or terminal end of the live wire the metal would become 'live'. With an earth wire the current will flow along it to earth, and the circuit fuse 'blow' so preventing a risk of electric shock. With table or standard lamps made of wood or pottery the risk is slight but, as copper, bronze and stainless steel are now attractively used to make cases, supports and shades, and as nearly all fluorescent fittings have metal cases, an effective earth is advisable. A good earth also assists fluorescent lamps to start reliably.

Appliances and lamps with electrical connections completely insulated from the user do not need an earth connection. In the case of appliances, these are usually marked 'double insulated' or have a symbol on them which shows a square inside a square. With lamps it is more difficult to tell if they are safely insulated but, as a general rule, all lamp fittings with metal stands or bases should have an earth wire and be connected to a circuit with an earth conductor. If in any doubt ask a qualified electrician or registered contractor.

The earth wire is just one of three wires common to all wiring. The difficulty here is that all the older wiring for lighting used to be made up of just two cables – the live and the neutral. For electricity flows only when it can complete a circuit – from the source of power along a wire (the conductor) and back to the source again. From the source it flows along the red sheathed wire called the *live* or *line* to the switch and from it to the lamp or the appliance and from it along the black sheathed wire called the *neutral.*

Colours for the flexible cord between lamp and plug are: brown for live, light blue for neutral and green/yellow for earth.

A short circuit occurs when electricity flows directly from the live to the neutral without passing through the lamp or the appliance.

Some of the older lighting wiring had the two cables quite separate – often rubber covered and drawn inside a steel pipe called a conduit. The conduit was a very good method in its day but it was costly, needed special tools and it has proved difficult to modernise. There are instances where aluminium and particularly high impact PVC conduit is

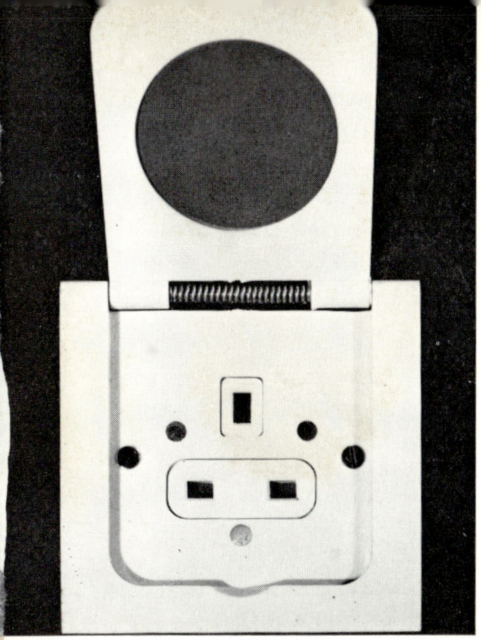

1 Hinge flap floor socket (Walsall Conduits Ltd)

2 Thyristor dimmer control (Nettle)

3 Fused Adaptor with four 5 amp two-pin plugs (Nettle)

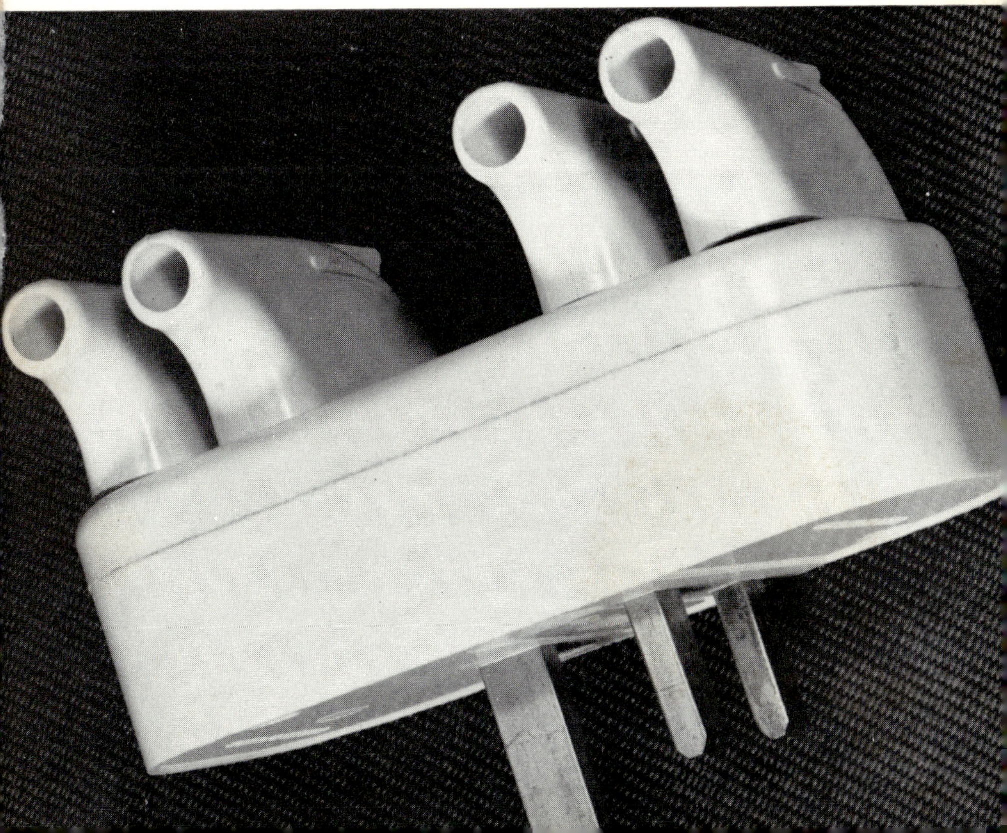

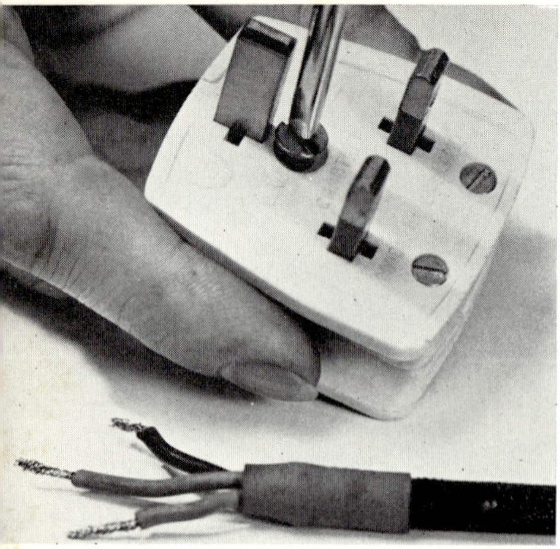

4

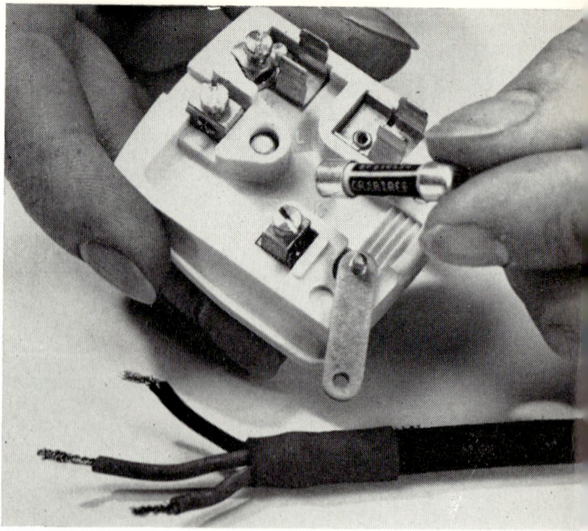

5

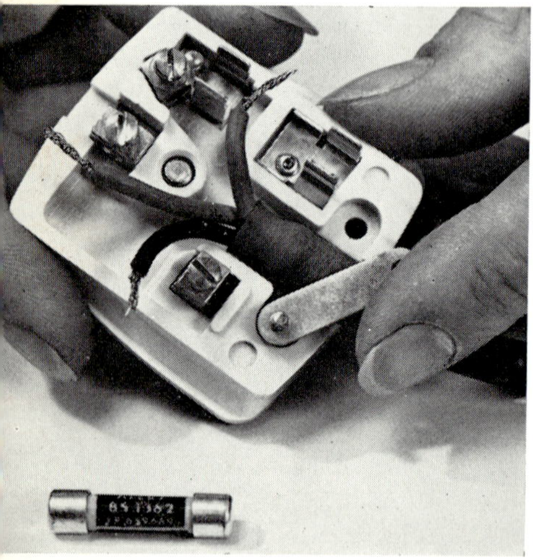

6

4–8 WIRING A PLUG. The full description of these pictures showing how to wire a plug are given opposite. These instructions must be carefully followed

7

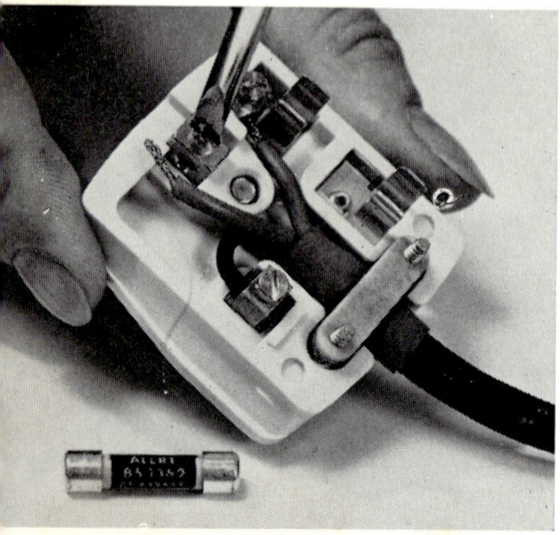

8

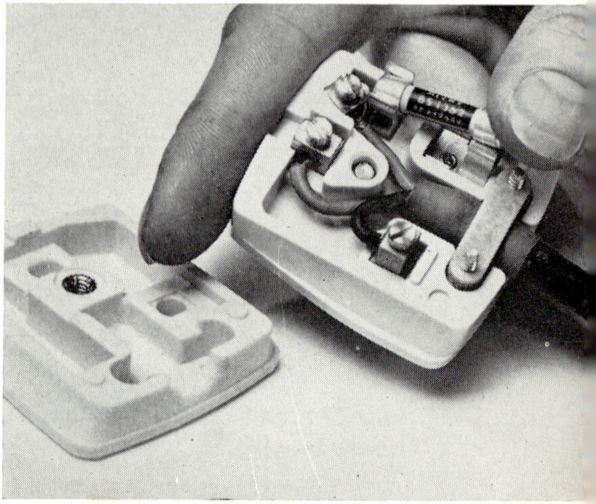

being used, but these days most wiring cables for the home are made in Poly-vinyl-chloride (PVC) which is tough, will not burn and is not affected by water, oil or most chemicals. It can safely be embedded in plaster, put under floorboards or in the ceiling – in fact, in most places in the home.

The cable now commonly used for lighting is 1 mm^2 PVC twin and earth, sheathed and rated for up to 12 amps. Prior to metrication it was 1.044 or 3.029. It is made up of two cores and earth – a red coloured insulated core for con-

HOW TO WIRE A PLUG

Wiring Method

1. First, with pins facing up, undo main screw to open. Loosen screws holding cord grip. (*Plate* 4, opposite)

2. Swing cord grip to one side. Take out cartridge fuse, unscrew three terminal screws. (*Plate* 5)

3. Make sure flex braid is stripped back 2 in., and the frayed ends bound with insulating tape or sealed with a rubber collar. Clamp taped portion under cord grip. Trim back wires – brown (red), blue (black), green/yellow (green) – to allow each to run to correct terminal, plus a portion for winding round or fitting into terminal. Strip about ½ in. of covering from the wire. (*Plate* 6)

4. Neaten bare wires. Wind brown (or red) – *Live* round fuse terminal, wind blue (or black) – *Neutral* round left-hand terminal, wind green yellow (or green) – *Earth* round top terminal. Screw back the terminal screws. On appliances where no earth wire is provided, attach the brown (or red) wire to the *Live* (L) terminal and blue (or black) to the *Neutral* (N). If both wires are the same colour (as sometimes happens with lighting flexes) connect one to the *Live* (L) and one to the *Neutral* (N). (*Plate* 7)

5. Push back fuse cartridge into clips, checking you have the right amp fuse for the appliance (3-amp fuse for lighting). Check again that you have connected up the correct colour to the correct terminal. Fit plug top back, screw up main screw. (*Plate* 8)

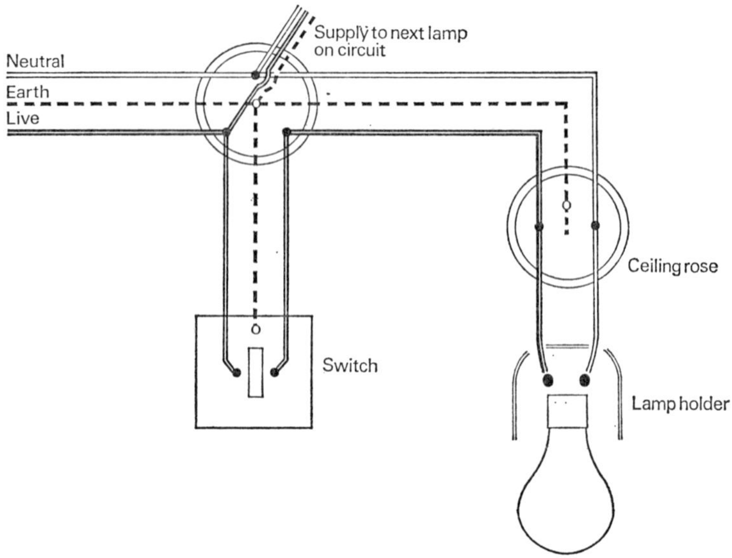

Fig. 1 Joint box method of wiring switch and lamp

nection for the live, and a black coloured insulated core for connection to the neutral, and between them a bare earth conductor wire. The three conductor wires are laid side by side, surrounded by a sheath – also in PVC.

Sometimes cables of a higher amperage are used, and there are specially made cables for outdoor use – particularly for garden lighting.

The lighting cable travels from the consumer unit to a series of lighting points. It has also to connect to the switch which turns the lamp on and off. It must do this in such a way that the individual switch (unless purposely designed) does not effect the operating of other lights. *Fig 1* shows the *Joint Box* method of wiring lights. With this method the lighting cable is taken to a series of joint boxes – as many as ten on the length of cable. At the joint box the live wire is taken down to the switch and back, then carried on to the ceiling rose. The earth wire also goes down to the switch.

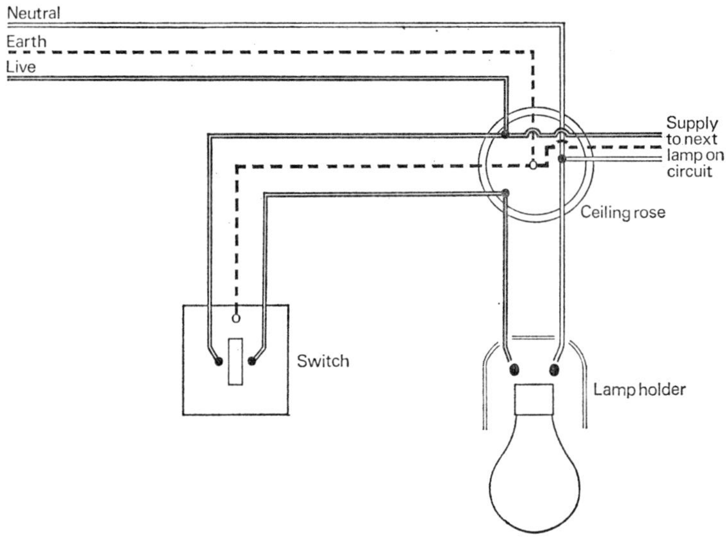

Fig. 2 Loop-in method of wiring switch and lamp

The wire is taken from the neutral and earth to the ceiling rose. A variation of the joint box method (not illustrated) has four terminals and an earth for connecting up a pair of lights.

Fig. 2 shows the more modern *Loop-in System.* The loop-in ceiling rose, *Fig. 3* (actually a combined joint box and ceiling rose) has four terminals: *live, neutral, switch wire* and *earthing* terminal. The neutral and switch wire terminals are connected to the flexible pendant cord. An advantage of the loop-in ceiling rose is that it is designed so that the various wiring connections can be easily made and the installation simplified.

With the *loop-in system* the wiring cable goes to each lighting point in turn. A separate cable goes from each lighting point to the switch controlling it. In the *joint box system* the twin core and earth cable is run to each joint box and a separate two-core and earth cable is run from the joint box

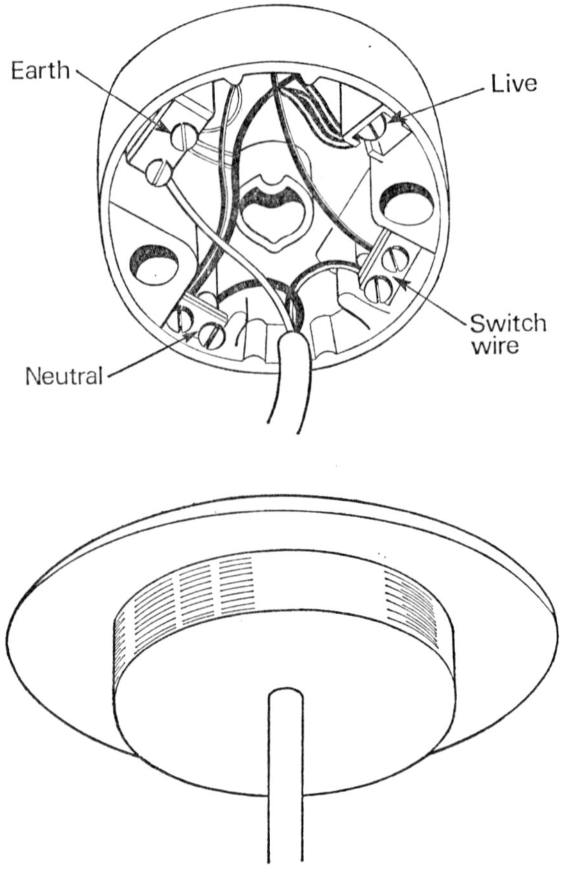

Fig. 3 Loop-in wiring in a ceiling rose

to the lighting point and another cable to the switch. The same type of loop-in ceiling roses can be used but without a live loop-in terminal.

Every home should have at least two lighting circuits protected by 5 amp fuses. Although a 5 amp fuse will carry the twelve 100 watt lamps (240 volts x 5 amps = 1 200 watts), it will be more realistic to plan one circuit for downstairs and one for upstairs.

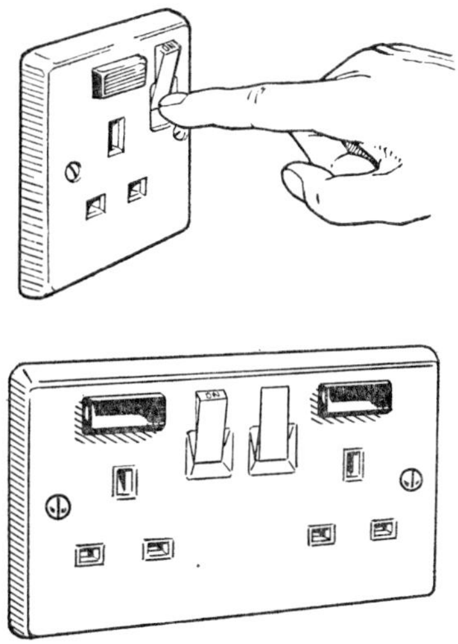

Fig. 4 13-amp socket outlets with switches that lie flush to surface when 'off', and red neon indicators (Crabtree)

THE RING CIRCUIT

Of course you may get some of your lighting from the socket outlets on the *ring circuit.*

Instead of a separate length of wiring going from the consumer unit to individual sockets a single loop of cable is put in to which any number of sockets can be connected. This is the modern *ring circuit* method of wiring. The cable used is 2.5 mm^2 PVC twin and earth sheathed. Total loading – the most electricity available from a ring circuit at any one time – is 7 200 watts (7.2kW). This is enough for all the most used home appliances as it is unlikely that everything will be on at once. An iron, a refrigerator and a television set on simultaneously would add up to just over 1 000 watts. Lamps and lighting use so little electricity that an almost unlimited number can be connected to the ring circuit either via a socket or direct to the fitting from a fused spur – a short length of cable going off like a branch line from the ring.

The ring circuit is protected from overloading by a 30

amp fuse and individual cartridge fuses in the individual square pin plugs.

INSTALLING LIGHTS

When a light is connected to a 13 amp socket outlet a twin or three-core circular sheathed flexible cord is used with a 3 amp fuse (coloured blue) in the plug. The other rating of cartridge fuse is the 13 amp (coloured brown) which is for appliances rated between 700 watts and 3 000 watts.

It is most important to remember that the new colours of the *flexible* cord between lamp and plug are: brown for live, light blue for neutral and green/yellow for earth. Previous to October 1970, when the new European standard came into use, the colours were: red for live, black for neutral and green for earth.

Floor standards, table lamps, bedhead lights, pin-up wall lights, dressing table lights, fixed and portable spotlights and permanent lighting may all be supplied from plugs and socket outlets using a flexible cord.

Where the lights are fixed and connected to a ring circuit a switched, fused spur unit instead of a plug and socket should be used – otherwise you have only the main ring circuit fuse to rely on. A 3 amp fuse should be fitted into the fuse carrier.

Pelmet lighting can be supplied from a fused spur unit (*Fig. 5*), so can fixed wall lights, and lights for a porch or garage or an awkward corner distant from a lighting circuit.

An additional light can be connected to the live, neutral and earth conductors of an existing lighting circuit at the loop-in ceiling rose, or the joint box. When there is no room for the new cable in the ceiling rose or joint box, a new joint box can be inserted into the feed cable. This is not always possible with a conduit wiring system and it may be best to run new wiring from the consumer unit.

Where you have no earth conductor in the existing circuit and the light fitting is made of metal – such as a fluorescent

Fig. 5 Fused switched spur can be used to connect a lighting point to the ring circuit (Crabtree)

lamp holder – then it is advisable to run new wiring or else utilise the ring circuit.

When the new light is to be controlled from an existing switch the electrician has only to run a two-core and earth sheathed cable from the existing light terminals to the new one. But if the light is to be turned on and off using a separate switch, then the electrician must wire to main, live and neutral (see *Figs. 1 and 2*).

The main difficulty is to add other extra lights with the minimum of disruption to decorations, walls and floors. Wiring is not beautiful and it needs to be disguised – also the PVC cable must be protected from damage. There are all sorts of ways of hiding wiring – the tricks of the trade – and your electrician should be able to advise you. There is, for instance, skirting, cornice and architrave made in a tough plastic that is designed to conceal cable, and it can be painted to match the woodwork.

PROFESSIONAL HELP

Get professional help for rewiring or for any scheme to improve your lighting. Be wary of odd-job electricians without adequate qualifications. Choose someone on the list of the National Inspection Council for Electrical Installation Contracting. You can find this list at your Electricity Board shop or in the public library. Many of the firms on the list will also be members of the Electrical Contractors' Association – a body who guarantee their members' work and also offer a personal loan scheme for householders who plan to do any major rewiring. Write to them at 55 Catherine Place, Westminster, if you need information or advice. Homes that were wired before 1939 should really be completely rewired – both lighting and socket outlets.

Standards for home wiring are established but prices are not. Members of the ECA and the Electricity Boards set their own prices but all must comply with the wiring regulations drawn up by the Institution of Electrical Engineers. Most libraries keep a copy or it can be purchased through your local bookseller.

Electrical contracting is a competitive business so get competitive quotes. As a very rough guide, calculate on £7 – £8 for each socket outlet; £6 – £7 for each lighting point; £12 – £14 for the consumer unit to control the circuits. Outside lighting points for porch and steps will cost a little more and there will be an additional charge for taking out the old wiring and making good the holes. Saving by cutting down on socket outlets can be false economy.

Next to safety, the best reason for rewiring the home is to improve the lighting and cure socket starvation. Wherever appropriate put in twin sockets as the difference in price is small. Choose switched sockets as they are safer and, again, the difference in cost is small.

SITING SOCKETS

Almost all the makers of sockets and switches now produce a true white instead of that unpleasant pale yellow or ivory colour that clashes with white paint. Switches and sockets designed with slim, neat finishes that look modern are widely available. Look at the catalogues and insist on having the ones you like. Most of the makers and all good contractors will send a consumer leaflet if you write to them. Note that telephone and TV aerial outlets are also available to match.

Site the sockets where they will be used. Avoid long flex runs. If you are going to use an appliance or a lamp in the centre of a room it may be worth while putting in a floor socket with a hinged cover. A floor socket can be very useful under a dining table for a lamp or for a standard lamp next to a favourite chair.

A twin socket outlet at the bedside allows you to have the blanket and the bedside lamp in use at the same time. There are, of course, adaptors but – if you have to use them – buy only those that contain their own fuse and are purpose designed. Extra socket systems of the do-it-yourself type should be treated with caution. They can prove more expensive than having an electrician do the job properly and some do not comply with the IEE wiring regulations.

Socket height depends on the type of house and its decorations – at least six inches above floor level and nine above a work surface is a guide. If you do not want the walls or floors dug into, then ask the electrician to show you samples of PVC skirting which replaces existing skirting and hides most of the cables.

LIGHT SWITCHES

The light switch has improved from the days of the half-apple shape with the small knob, although there are those who say that it was the only really *beautiful* switch with its circular porcelain base and decorated brass cap. True, they were efficient and robust but – unless you hanker for Edwardiana – they really do not match modern furnishings.

The modern switch is either of the *rocker* or the *dolly* type, which refers to the switch action rather than being a social comment! The most common now is the rocker which needs only slight pressure from a finger to switch the light on or off (*Fig 6*), while the dolly is a firm, up and down

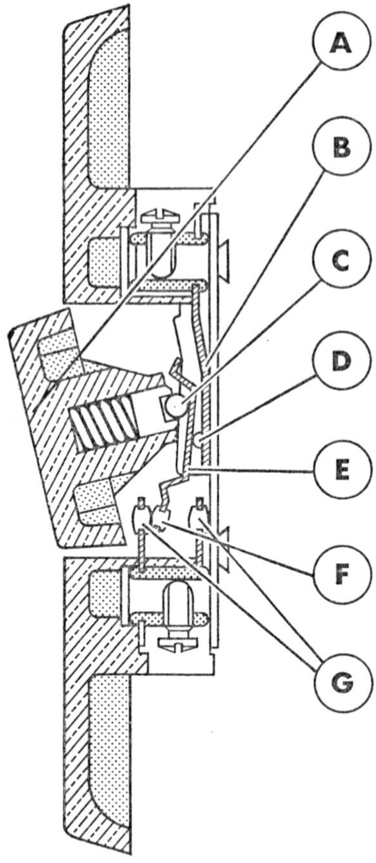

Fig. 6 Section view of a two-way switch. Finger pressure on the rocker (A) resting on the spring-loaded steel ball (C) moves the current carrying switch contact (E) with its double head (F) to engage alternately with each of the fixed contact studs (G). Current passes then down (B) through the pivot point (D) to reach one or other of the points (G) which leads to the twin wire joining the matching two-way switch in the circuit (see Fig. 12)

action. Most switches are of the flush type (*Fig. 7*), that is the mechanical part is recessed into the wall. If it is not convenient to cut into the wall, however, the switch can be

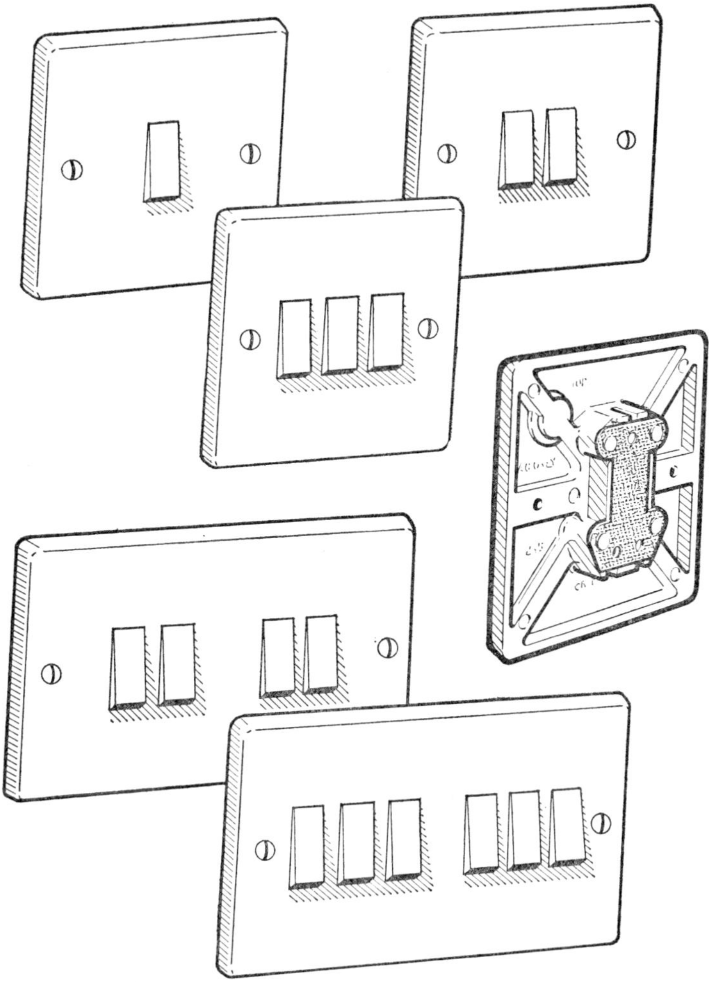

Fig. 7 Modern rocker switches are slim and need only slight finger pressure to switch on or off. Up to six lights can be controlled from one position (Crabtree)

surface-mounted on a shallow box or a pattress, that matches the switch.

The metal, plaster-depth box for flush switches normally measures 73 mm wide by 73 mm high by 28 mm deep. The surface plate can take up to three switches in line. Up to

Fig. 8 Architrave 1-2 gang switches and boxes (Crabtree)

six switches can be mounted on the larger plaster-depth boxes measuring 133.35 mm wide by 73 mm high.

A problem with the flush switches is that, after a time, a patch of grease or dirt appears on the wallpaper round the switch. To avoid this and to provide a neat fitting the *architrave switch* (*Fig. 8*) is now coming into general use. This flush switch is inserted in the architrave of the door frame which can be quite easily removed and replaced. The cable feeding the switch can then be run down the channel behind the architrave. There is a specially designed PVC architrave moulded to take the flush switches which are mounted one above the other. They are small and neat; overall size of a twin switch is only 120.6 mm tall by 32 mm wide and protrudes 9.5 mm from the surface.

In the jargon of the electrician a twin switch is known as

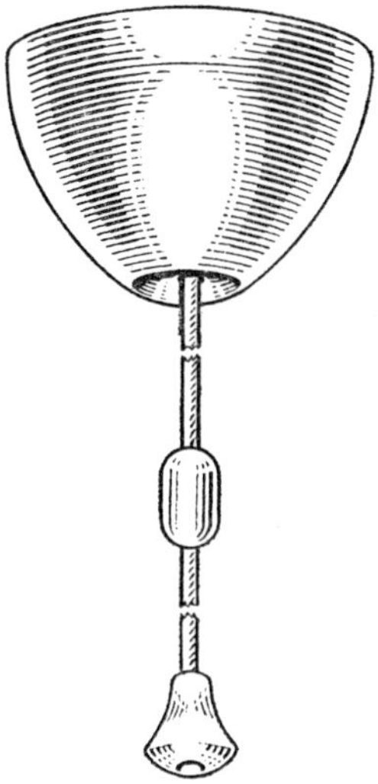

Fig. 9 Ceiling switch (Crabtree)

a '2-gang', a single switch as a '1-gang' – worth remembering when you start looking at the catalogues.

Then there is the *ceiling switch* (*Fig. 9*) noteworthy because of its safety. Operated by a shock-proof cord, the switch and the wiring are out of reach. A pull from almost any angle switches on or off and it can be used, not just for lighting, but for permanently fixed appliances – particularly heaters. It is the only type of switch permitted in the bathroom.

Yet another kind is the *dimming switch* – a means of controlling the intensity of the lighting – described in more detail on page 83.

You can also buy switches that have a red neon indicator – useful as a reminder where they operate a light source distant from the switch (see *Fig. 4*.) There are also switches with a time delay – they switch themselves off automatically after a short period. They are frequently used on common staircases in older flats, a practice to be deplored as – apart from displaying the avarice of the landlord – it is dangerous for older people who may not possess the alacrity to reach the top before the light goes out.

SWITCH CONTROL

Your electrician will help you to site your switches but it is useful to know some of the ways that switches can be used.

One Way Wiring. Fig. 10 shows how the ordinary *single pole* switch is connected. Single pole means it is connected in circuit with the live conductor. In this and the next seven figures the earth connection has been omitted to make it easier to understand.

Fig. 10 One way switching

Sometimes a *double pole switch* is fitted. With this kind, when in the 'off' position, both live and neutral conductors are completely isolated from the supply. This type of switch is intended for use in situations where complete isolation of the circuit is required in an emergency or for maintenance purposes. The principal purpose of a double pole switch is for the isolation of sub-circuits, the local control of cookers and immersion heaters. The standard method of connection is shown in *Fig. 11*.

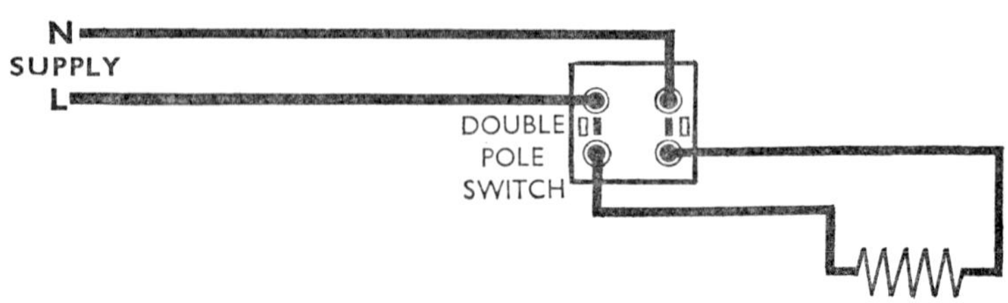

Fig. 11 Double pole switching

Fig. 12 Two way switching

Two Way Switching. The two way switch is the one that allows you to switch on and off from two different positions. This type should be installed in bedrooms, on landings and in long halls, on single-flight stairways and in any room with two doors. *Fig. 12* shows the generally accepted method of wiring two way circuits.

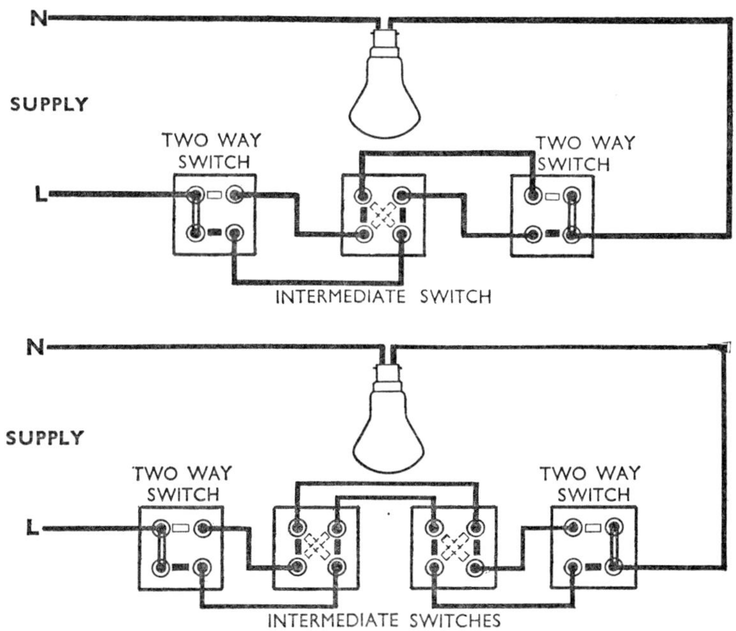

Figs. 13 & 14 Intermediate switching (1) and intermediate switching (2)

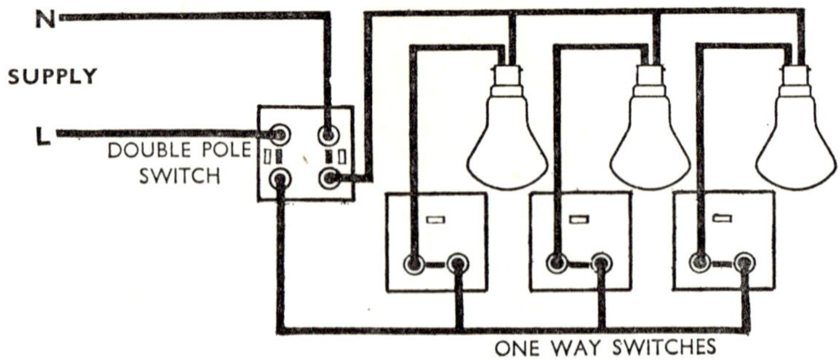

Fig. 15 Master switching (1)

Intermediate Switching. Long halls, corridors and passageways with many doors, and multi-flight stairways are still often wired only for two way control when – for safety and convenience – every access point should have its own lighting control switch. This is not difficult or expensive if it is done when rewiring and any number of switches can be introduced into a two way circuit. *Fig. 13* shows how a single intermediate switch should be connected when this is to be added to a two way circuit. The method of connecting two or more switches is illustrated in *Fig. 14.*

Master Switching. When a master switch is mounted near the main door to a house or flat where it is an easy means of controlling a number of lights from one position – very useful when going out or returning at night.

Fig. 15 shows a lighting circuit which each light individually controlled by a single pole switch, but with a double pole switch as the master control.

Another way of doing it is shown in *Fig. 16.* This method is used when you want to provide a policeman or nightwatchman with a means of momentarily flooding certain areas with light without having to operate a number of separate switches. In this circuit the lights are separately controlled by two way switches, wired as one way switches. The third terminals are coupled to a single pole master control device.

9 Slimline fluorescent (three straw texture diffusers mounted together to prevent lamp being seen from sides) lights alcove (Atlas). Watercolour by Samuel Prout is lit by spotlamp mounted on opposite wall (Rotaflex)

10 Table lamp with its own dimmer (AEI)

11 Lighting a picture using a 284 mm tubular filament lamp in a modern fitting (H.H. Electrics)

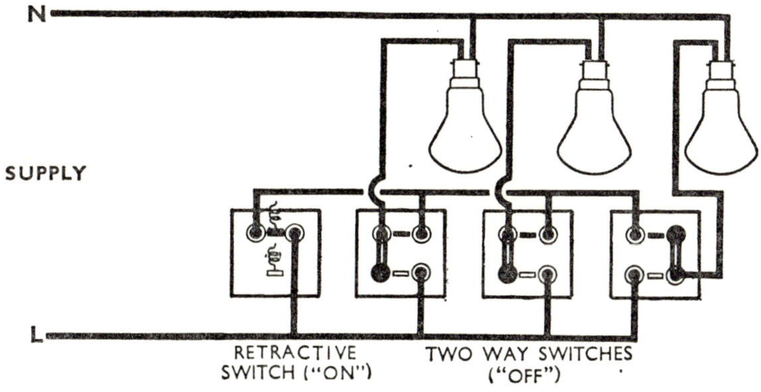

Fig. 16 Master switching (2)

This can be a switch or bell push with a retractive action – that is, a switch that works only so long as a finger holds it in the 'on' position.

Fig. 17 shows a variation of the idea – a one way switch has been placed into a two way circuit so as to prevent the lights being extinguished from either of the two way switches. This system is not really for the home but the lighting circuits for stairways of large hotels, restaurants and guest houses where accidentally turning off the stair lights could be dangerous and complete two way control is required by the

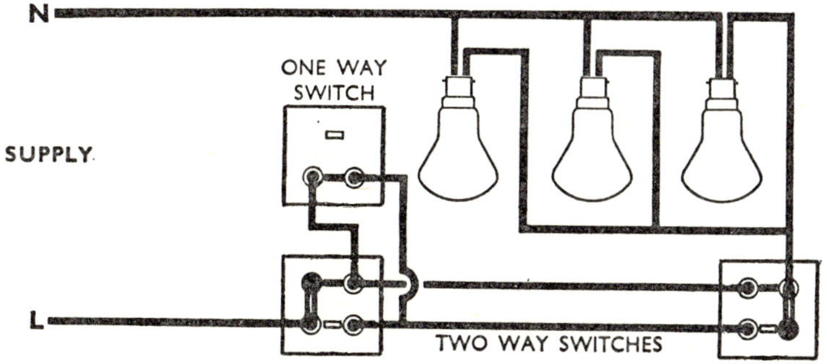

Fig. 17 Master switching (3)

staff when the premises are closed to the public or during the daytime.

In the home these methods of master switching can be used as a burglar deterent or can be linked to an intruder alarm system – so all the lights go on automatically if someone trips or activates an alarm switch. Another idea is to connect a group of lights to a master control circuit and have the electrician fit a photo-electric (light sensitive) switch which turns on the lights when darkness falls and turns them off again at daybreak. It gives the impression that the house is occupied, which can be useful during holiday periods. These light sensitive switches are not expensive or difficult to install, although positioning is important – the switch must not be installed in direct sunlight, for instance. If you do not want the light left on all night then the same result can be obtained with the time switch set to turn off at midnight.

Photo-electric beams and sound or heat sensitive switches can also be fitted but they are usually not cheap and their reliability is sometimes questionable. You can fit a switch under a carpet or a door mat that is pressure sensitive so that, as you enter a room, the light will go on but, again, use is limited as any movement of traffic over the spot will switch them off – probably inconveniently.

Before even thinking of installing this kind of sophisticated control, get professional advice from an electrician who has seen or used them.

LIGHTING TRACK

The best way of getting more light without rewiring or tearing into walls and ceiling is to use electrified track. (*Fig. 18*) The extruded aluminium track is wired from a single point, and a variety of light fittings can be clipped into the track anywhere along its length. The track can be recessed into the wall or ceiling or mounted on the surface or suspended from pendants. (*Fig. 19*) Although it is a con-

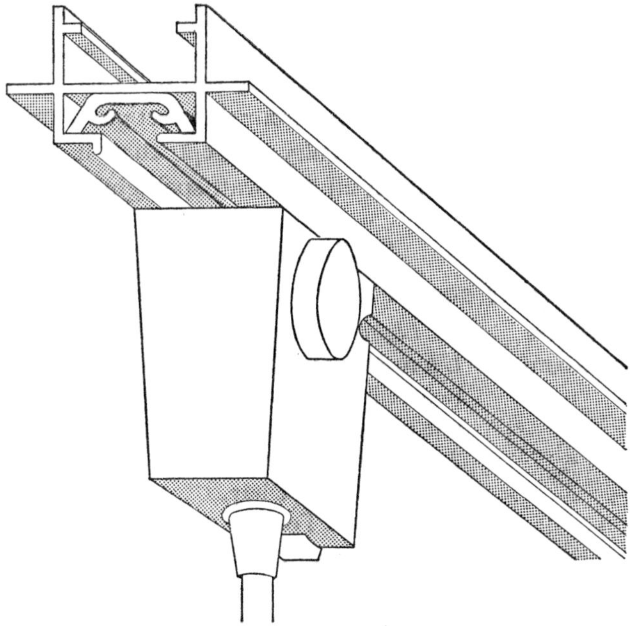

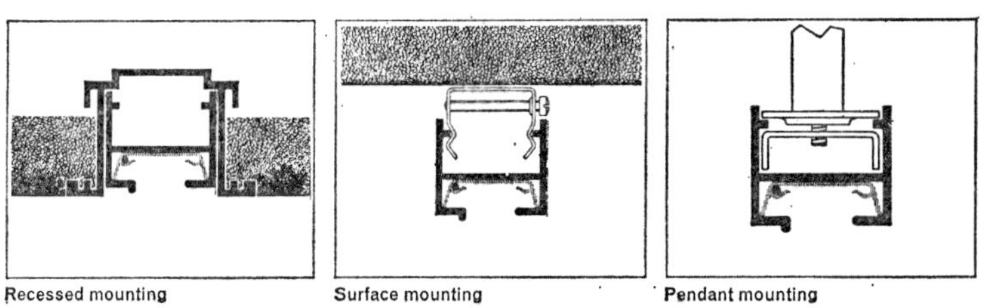

Figs. 18 & 19 Electrified tracks

tinuous electrical power source, the conductors are safely concealed out of touch.

Standard lengths are 1.2 m, 1.8 m, and 2.7 m. Lengths can be joined to form any pattern. Maximum load for an individual track run connected to a power supply is 16 amps (3.75 kW at 240V) equivalent to 36 x 100W or 24 x 150W or 12 x 300W spotlights.

If you use a section of track in place of a centre ceiling

point – a simple solution to the centre fixture problem – then you are limited by the load factor of the 5 amp lighting circuit – 1 200 watts. As it is unlikely you will need more than four or five lights in the room from that section of track, you should have no load difficulty – except that it would be inadvisable to make them all 300 watt spotlight fittings.

TABLE ONE

SUFFICIENT SOCKET OUTLETS AND LIGHTING POINTS

Where	*Lighting Points*	*Socket Outlets*
Living Room	*Up to 15 m^2:* One or two ceiling points, one or two on each of 2 walls, a point in the main window or bay.	4 double, one on each wall near a corner
	Over 15 m^2: For square rooms – between one and four ceiling points or four wall points, where length exceeds width, at least two ceiling or four wall points.	5 double, two on 'focal' or fireplace wall and one on each other wall
Dining Room	A point over the table	2 double, not on same wall, 1 floor outlet by table
Kitchen	A point on the wall or an adjacent ceiling area for each wall where there is a working surface In light coloured kitchen up to $11m^2$ two fluorescent fittings on the ceiling will give good lighting	A double above each working surface, in addition to singles for washing machine, refrigerator and dishwashers (Permanent appliances wired to switched spurs or sockets). A separate ring circuit is recommended for the kitchen

Where	*Lighting Points*	*Socket Outlets*
Double Bedroom	1 ceiling and 1 wall point for dressing table area and bed-head. Ideally, two pendant drops – one each side of mirror	3 double, one on each of three walls preferably each side of bedhead and one on facing wall
Single Bedroom	1 ceiling, 1 or 2 wall points	2 double, one next to bed
Study/Bedroom	1 ceiling, 1 or 2 wall points	2 double, one next to bed, one for desk
Hall or Landing	1 point on each, so placed so as to light whole staircase	2 – a single on each, plus one on a mezzanine landing
Workshop	1 ceiling, over bench position	2 double, one floor, one above bench
Garage	1 ceiling, near far end from outside door, on driver's side	2 single, one at floor, one above a bench. (Frequently a food freezer is sited in a garage).
Bathroom	*Below 5 m^2:* 1 ceiling *Over 5 m^2:* 1 ceiling and 1 wall over mirror position	Supply for heater or towel rail (Not socket outlet). Point for shaver socket
WC/Cloakroom	1 ceiling, 1 wall over mirror position	
Utility Room	1 ceiling	1 double over work-top, one double above splash line
Attic/Loft	1 ceiling	1 single
Front Door	1 exterior	
Back Door	1 exterior	
Patio	1 ceiling or wall	1 (weatherproof)

A length of four foot track costs under £6 and the adaptors that fit into the track are about £1.75 each. You will need cord pull switches or switches on the fittings to allow you to vary the lighting. There is also double circuit track which will allow two sets of lights to be switched on independently or together.

Lighting track is widely used by shops to improve their lighting and it has the advantage that the light fittings can easily be moved to suit changes in the display. Often it is used as the support/divider for illuminated ceilings – an idea which could be adapted to the kitchen.

Chapter 2: LIVING WITH LIGHT

To talk about *artificial* light is misleading. There are many kinds of electric light that are indistinguishable from the light one would get on a good clear day in summer. Indeed, electric light provides a better standard of lighting since there are constant variations in normal daylight.

The healthy human eye, of course, is designed marvellously to adapt to all levels of lighting. In some ways it works like the camera lens. Both are normally focused at infinity so there has to be some means of adjustment or all we would see at close range would be an indistinct blur. The camera lens is mechanically adjusted – the eye has muscles to do the job automatically. The pupil of the eye – like the retina of the camera – can also adjust to control the amount of light that enters. But as we get older the eye muscles and the retina work less efficiently. At sixty years more than twice as much light is needed as at forty. Adjusting from semi-darkness to bright light is never almost instantaneous action, which is why a good, even level of illumination is needed in the home for older people. This is particularly important on stairs, in corridors and in bathrooms and toilets where a great many falling accidents occur.

Spectacles can usually correct defects in vision in both young and old but it is still important not to overwork the eyes. Standards are now applied to lighting at work to the benefit of physical and mental health so it is equally important to achieve the same standards at home.

MEASURING LIGHT

To understand what light is means going into the technical

detail – a scientific subject outside the scope of this book. The simple facts are these: white light is a mixture of the colours of the rainbow. The impression of colour understood by the brain will be different on a sunny day to that on a dull or misty day. Too much light directed to the eye is harmful. Explorers in the polar regions will tell you that light reflected from white snow and ice can cause nausea, dizziness and depression. Walking or driving directly into the rising or setting sun is physically unpleasant and the popular sale of polarized glasses designed to cut down daylight glare emphasises the point that our appreciation of light is rather a subjective business. The rather curious, old-fashioned notion that 'artificial light is bad for the eyes' is very difficult to understand when you start to think about it.

If you are interested in the scientific side there are numerous books on the subject. For our purposes, it is only useful to know how light output is measured. This definition is taken from the Interior Lighting Design Handbook published by the Lighting Industry Federation and the Electricity Council.

The illumination of a surface is measured by the concentration of light falling on it. To express it in numbers we need a unit of light – the lumen (lm). So we can say that the output of a 100W lamp is 1 260 lumens and a 60W lamp 665 lumens, or that the recommended minimum illumination for a particular room is so-many lumens per unit area.

In imperial units it was measured in lumens per square foot (lm/ft^2), at one time called 'foot-candles' – a term still used in the U.S.A. In metric units illumination is measured in lumens per square metre, or 'lux'. Since a metre is larger than a foot, one lumen spread over a square metre is a smaller illumination than the same amount of light over a square foot. The relationship between the two units is thus the ratio between a square metre and a square foot.

Conversion is simple: there are 10.76 square feet to one square metre so $1\ lm/ft^2 = 10.76$ lux. Illumination from natural sources varies from 0.1 lux, or less in moonlight, to as much as 100,000 lux in direct sunlight in the tropics – a range of 1,000,000 to 1. A typical figure for the illumination out of doors from an unobstructed overcast sky in Great Britain is 5,000 lux (nearly 500 $1m/ft^2$); this is adopted as the basis for daylighting calculations in architecture.

Lighting in buildings varies from about 1 200 lux down to 100 lux or less – though the Illuminating Engineering Society's recommendations for areas where people are engaged in continuous work (as distinct from 'passage' areas) are in general not lower than 200 lux, which they consider to be the minimum amenity level for work places.

If we take the light output of the lamps used and multiply it by their number the product represents the light that has been 'put into' the lighting scheme. Take a room lit by eight 100W lamps and ten 60W lamps. The total 'installed light' would be:

$$8(1\,260)+10(665)=16\,730\ \text{1m.}$$

Strictly speaking, since light is emitted continuously from a source in all directions and is steadily arriving at an illuminated surface, it is not the total quantity of light the illuminating engineer is normally concerned with but the rate at which it is being given out. Thus the lumen is not really a unit of light itself but rather one of its flow. This is similar to the situation with electric current. For, as explained in Chapter 1, the ampere (amp) is not a unit of *quantity* of electricity but of its flow. If we want a real unit of quantity of light we must multiply lumens by time and speak of 'lumen-seconds' or 'lumen-hours'. Normally, however, it is the flow or 'flux' – the quantity per unit time – which is important. Formal definitions of the lumen speak of it as the unit of 'luminous flux' or installed flux per square metre'. But, for our purposes, it is easier to talk in terms lux and Table 2 gives the levels of illumination recommended in the home in these terms.

Compare these recommendations for the home with the

minimum recommendations for general office lighting. In Britain it is 400 lux, in Europe it is 500 lux but in America it is 1 076 lux.

TABLE TWO

RECOMMENDED ILLUMINATION LEVELS

Kitchens	200 lux
Stairs	100 lux
Workshops	400 lux
Garages	50 lux
Living Rooms	100 lux
Sewing and Darning	600 lux
Reading	200 lux
Homework	400 lux
Bathrooms	100 lux
Bedrooms	50 lux
Bedside	200 lux

It is common knowledge, backed by international scientific evidence that, under better illumination, seeing becomes faster, more accurate and involves less fatigue. So lighting is a tool for efficiency and productivity, as well as being necessary for welfare, safety and aesthetic harmony.

AVOIDING GLARE

Lighting is a matter of quality as well as quantity. The type of lighting is very important for, if you have too high a contrast, you will get *glare*. In nature you can get glare direct from sunlight reflected from water or snow. In buildings you can get glare from windows and from light sources – either direct from the source to the eye or reflected from light coloured surfaces. If the background is much brighter than what we are looking at it can often cause eye strain.

Strain begins when the eyes try to cope with a too great variation in brightness. Different colours and textures in furnishings will absorb varying amounts of light – a dark fitted carpet will absorb more than a vinyl tiled floor; so will a dark wood desk and study furnishings. The Victorian panelled study lit only by a single lamp has great evocative charm but it has no place in a modern home.

Choice of the right type of lighting sustains attention and endurance. It also affects your sense of comfort in relation to surroundings and this has a direct effect on concentration.

Children's homework or any kind of sustained reading needs an illumination level of 400 lux and if the eyes are asked to cope over long periods of time with inadequate light and – at the same time – do difficult tasks like reading pages of small print, they soon organise physical protest meetings.

Researchers have found that the memorising ability of the brain is reduced under poor lighting or where there is harsh glare. Fatigue follows and then headaches.

Every home needs a good general level of lighting. Recommended levels of light for different rooms are given in *Table two* but it is probably easier to use as a general guide a recommended standard of 25 watts of lighting for every square metre. The standard was formally between 1½ and 2 watts for each square foot of floor space. Standard lighting easy on the eyes but lacking in interest – like the lighting in the factory canteen.

This is where local lamps come in, giving form and shading which helps us to recognise depth. Without it all objects look dull and uninteresting. But the shading must be controlled; those long, hard shadows that distort shapes are worse than none.

Test this for yourself by using a pen under a desk lamp placed close to the paper and watch the shape of the shadows. Print several large capitals first with the shadow of your hand going one way. Now try an exact repeat with the shadow coming from the other side. Not very easy.

Writing or reading in an armchair needs good local lighting. A 1.2 m standard lamp beside the chair, with two 60 watt lamps will give about 200 lux on the book, which is about adequate. Move the lamp 1.2 m away and the lighting drops to only 60 lux – causing eyestrain. If the standard lamp, desk lamp or spotlight is the only light in the room, then the brightness of the white pages of the book will contrast too strongly with the dark background. This causes glare. It can also send you to sleep – as anyone can testify who has watched home movies or slides projected in a darkened, warm room.

The aim is to get the right balance between diffused and directional light. Generally speaking, it is almost impossible to have too much light in a room – only too much shadow or unpleasant glare.

A good standard of lighting for a living room measuring 6 m by 4.5 m would be four spaced ceiling fittings – each with a 100 watt lamp, a 60 watt wall bracket or small table lamp at the desk; a 150 watt lamp in a floor-standard and 40 watt table lamp.

You also need to include lighting that will give a degree of sparkle; something that will enliven the whole scene – rather like the garnish on a dish.

The kind of light you get is far more important than the specific fitting. Part of the art of lighting is knowing what lamps to use to create a feeling of space. Light directed at the ceiling creates an illusion of height; light directed on the floor darkens and tends to lower the ceiling; a narrow room can be broadened by concentrating on lighting the two narrow walls and leaving the larger walls in comparative darkness; and lighting an alcove or a bay window makes it appear larger and more interesting.

LIGHT SOURCES

The gas mantle is an example of light from incandescence. Combustion of the gas provides the heat which raises the material of the mantle to a temperature which gives off white light. For it is a property of matter that, when it is

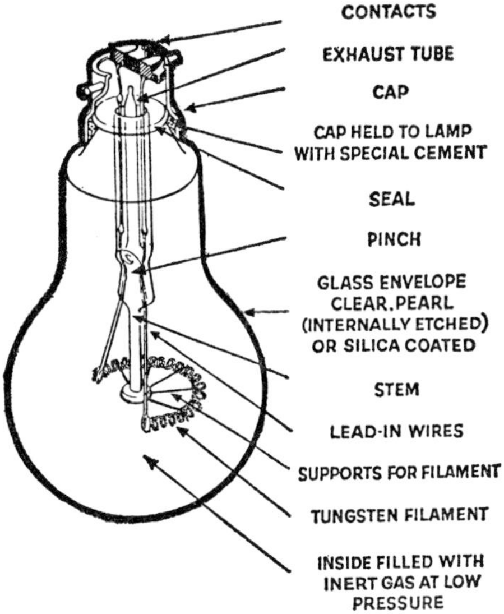

Fig. 20 Coiled-coil electric lamp showing the components

heated, it emits light – first a dull red then progressively whiter as the temperature rises. Most substances undergo a radical chemical change before incandescence is reached so that the number of materials that can be used as a light source is limited.

When electric current is passed through a wire conductor the resistance to its passage produces heat, and when a temperature of 500°C is reached the conductor – called the filament – emits light. This proved to be the first practical, simple form of electric lighting. The early difficulty was that the carbon or metal filament used in the lamp was destroyed by the oxygen in the air. The solution found in 1845 was to draw all the air out of the glass lamp – in other words, create a vacuum. By 1879 Edison in the United States and Swan in England were marketing vacuum lamps that gave more light than anything of similar size – and began a revolution in the way we live and see.

Filament (Incandescent) Lamps. Today the filaments are of tungsten formed into a coiled spiral. In place of the vacuum the lamps are filled with an inert gas, such as argon – nitrogen. This gas prevents the evaporation of molecules of tungsten from the surface and so higher operating temperatures are possible – which means more light.

The filament is coiled once or twice. The twice coiled type, called coiled coil, gives off more light than single coil and is slightly more expensive but overall the light is cheaper. The filament has an operating temperature of about 2 700K.

Temperatures are now calculated in Kelvin units (K) in the electrical and construction industries but the K unit is the same as the C degree unit.

Special short life lamps for photographers' use can be designed to operate at temperatures around 3 000K and higher but life is limited to about eight hours against the 1,000 hours of the average lamp.

There are filament lamps that work at a lower temperature and so have an extended life – about 2 000 hours. These are called *long life lamps*. They cost about 20 per cent more than normal lamps but have a lower light output (around 7½ per cent less light) for the same electricity consumption. They do not need changing so often which may be helpful in situations where a lighting fitting is installed in an inaccessible position. When choosing which type of lamp to buy it should be remembered that when lighting a room the major proportion of the cost is for the electricity consumed, the lamp itself representing only about 10 per cent of the total. In other words, you pay more for your light by using 'long life' lamps, but you may consider the extra cost worthwhile because they do not have to be renewed so often.

Most filament lamps used around the house are 40, 60, 100, 150 and 200 watts – all with bayonet (BC) caps and part of a range known as general lighting service (GLS). These lamps go up to 500 watts.

The filament lamp commonly used in the home comes in

a variety of finishes: *clear*, for reflector fittings and glass chandeliers, *pearl* (which gives a softer glow, kinder to eyes) for ordinary fittings, or *silica-coated* (giving even, diffused light) for fittings where the lamps show and you want to cut down glare as much as possible. There are places where you can use a colour lamp such as *pink pearl* or *amber* or *yellow* which give a pale colour light suitable for parties or background lighting.

With a *clear* lamp the filament is visible and so the shadows cast by objects close by are sharp and hard. With *pearl* lamps the inside is roughened by an acid treatment and there are no hard shadow lines and virtually no loss of light output. Softest shadows are produced by the *silica-coated* – a white powder, commonly titanium dioxide, which spreads the light with an almost uniform brightness. To avoid glare, use a silica coated or pearl lamp in conventional filament lamp fittings in the home.

Filament lamps are either pear shaped or mushroom shaped – the mushroom shaped are smaller, neater to handle.

Never put into a light fitting a higher wattage filament lamp than the wattage recommended – this is very tempting with mushroom lamps which are smaller by comparison for a given wattage. Correct wattage should be marked on the light fitting. Too powerful a lamp can damage the shade or fitting by overheating.

Make sure your lamp is of the same voltage as the mains in your home. If your supply is 240 volts and you are sold a 230 volts lamp, you will get more light but the life of the lamp will be much shorter. You will get less light and longer life for the lamp if the situation is the other way round – 240 volt lamp on 230 volt mains.

Filament lamps are also made in a great variety of shapes and sizes to suit special applications and fittings. *Candle lamps* and *round bulb* are recommended for fittings where the lamp is on display as part of the decorative design. *Tubular filaments* are used in picture lights or where a fluorescent would give too much light. There are also special filament lamps for cookers, fuel effect fires, indicators and pigmy lamps for a variety of uses like lighting Christmas trees.

Nightlight lamps, either 5 or 15 watts, are used to give a reassuring glow in a nursery or sickroom and these use hardly any electricity.

Remember always to keep spare lamps in a cupboard close to hand. Restock at regular intervals, otherwise you will find yourself at some inconvenient moment without a lamp of the correct type and wattage to replace one that has died.

Reflector lamps have their own built in reflector surface behind the filament. Some are shaped to give wide beams for floodlighting and others give narrow circular or oval beams for spotlighting. The two main kinds, ISL (internally silvered lamp) and PAR (parabolic aluminized reflector) (*Fig. 21*) are shown in Table 3. PAR lamps have a longer life – about 1,500 hours – and are tough enough to use out of doors. A difficulty with this type of lamp is the heat output which can cause problems in some situations but here the *dichroic* type can be used which gives a cool concentrated light. They are available in a choice of colours. Two or three fittings with different colour reflector lamps can be used to create wonderful effects.

Low voltage reflector lamps give a compact concentration of light from a compact sized fitting. At 12 or 24 volts, the filament can be much shorter and fatter than at mains voltage. A transformer is needed for these lamps, which costs extra, but some fittings have their own built-in.

Tungsten Halogen lamps give a light of a very high intensity from a very small glass envelope. The gas in the lamp has a trace of a halogen such as iodine in it and this

12 Tungsten halogen spot lamp designed to produce exact light shapes (Concord Rotaflex)
13 Only 190 mm wide this small fitting has a 500W tungsten halogen lamp that can be used indoors or outdoors for floodlighting (Concord Rotaflex)

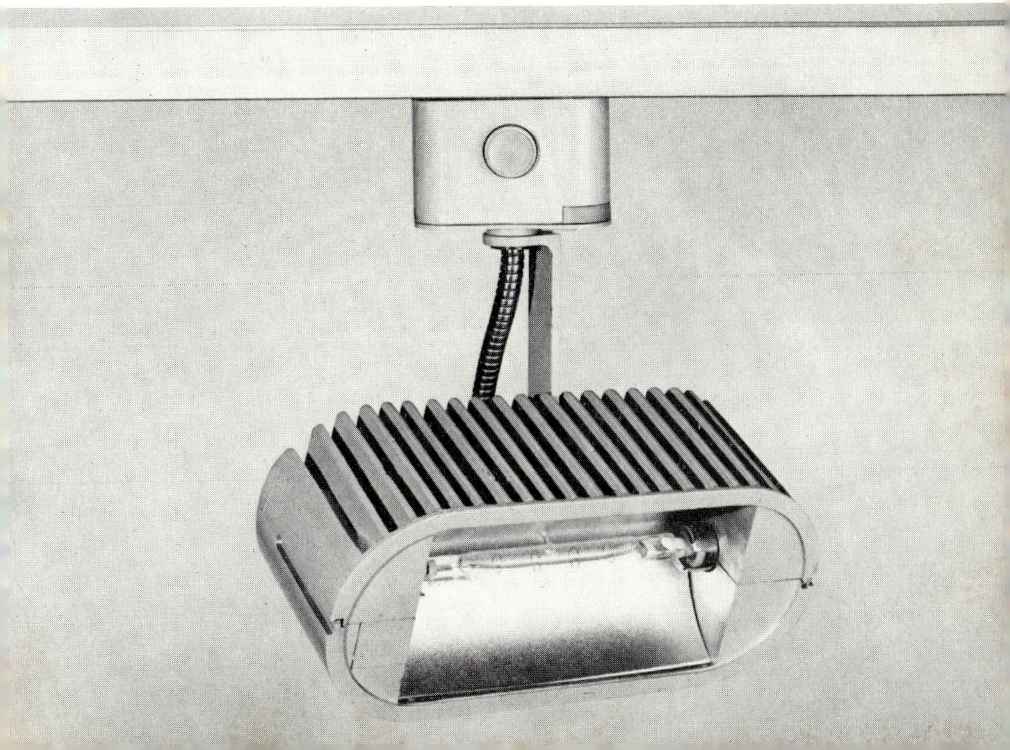

14 Opal glass fitting (light brick) used for house identification (Lumitron)

15 Adjustable wall light swings on double jointed arm (Conelight)

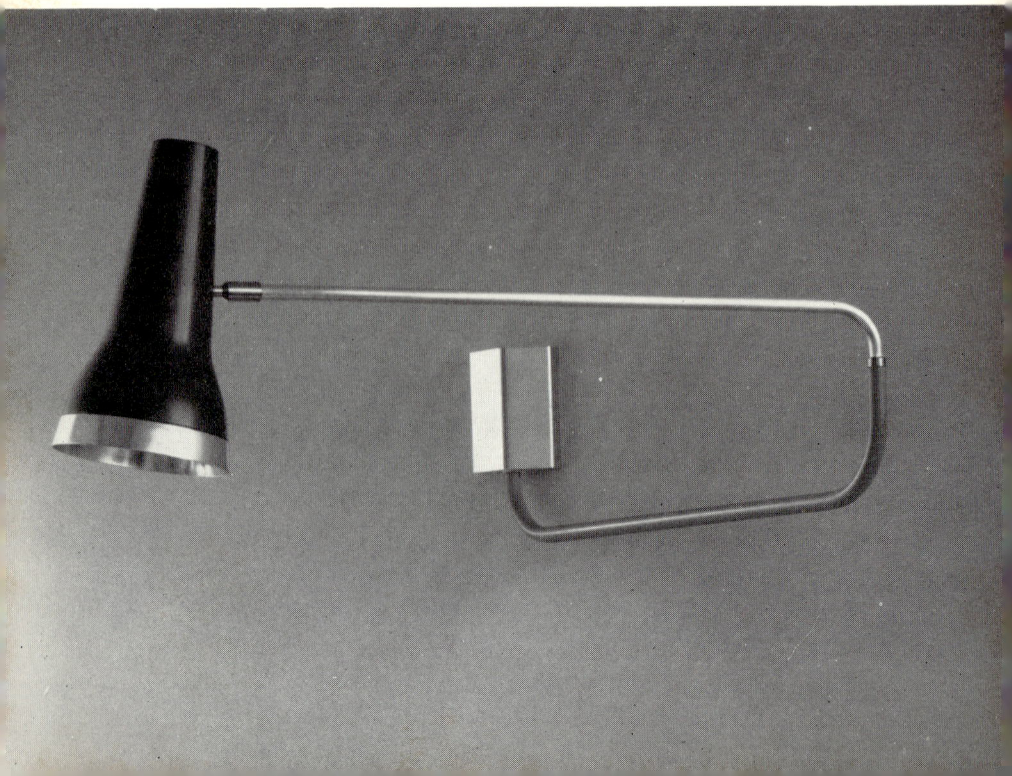

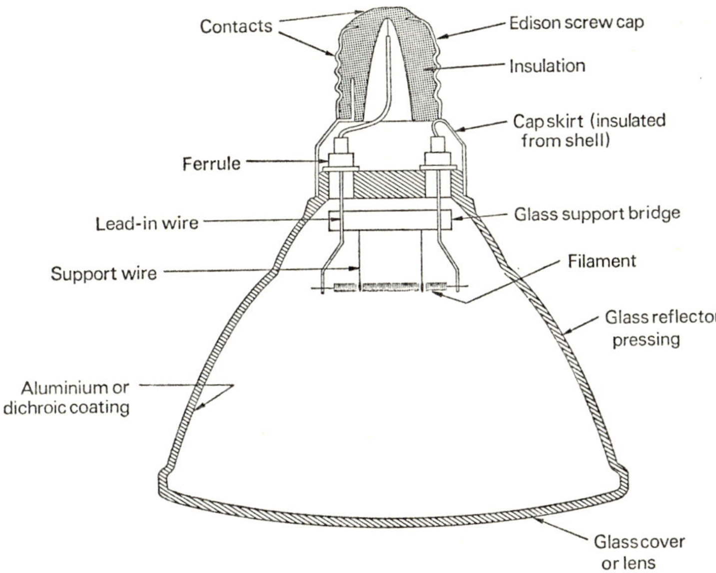

Fig. 21 Parabolic Aluminized Reflector (PAR) lamp showing the components

prevents blackening and has a restorative effect on the tungsten filament, extending its life and increasing its efficacy. This type of lamp is often used in movie and slide projectors but, because of their compact size, they are now being fitted in some low voltage spotlamps for the home and are likely to be used more and more. A higher wattage can be obtained in a much smaller size than with a GLS lamp, and they are more efficient and last longer. Life of a 12 volt, 50W tungsten halogen lamp is about 2 000 hours.

A *fluorescent lamp* is a glass tube coated inside with a special powder which gives off light. It has no filament at all. Instead, there are two 'cathodes' – one at each end of the tube – and the lamp is filled with a small amount of mercury gas which 'glows' when the lamp is switched on. There is not much light from the glowing gas but it produces quite a lot of ultra-voilet light which is invisible. *Violet* light is at the

TABLE THREE

LAMPS FOR THE HOME

Some of Principal Types on Mains Voltage (*see Fig. 25, page 57*)

Type		*Wattage*
Pear Shape Filament (Single and coiled) GLS	Clear	40, 60, 100, 150, 200
	Clear*	200
	Silica and Pearl	40, 60, 100, 150
	Daylight Blue	40, 60, 100
	Pink and coloured	15, 25, 40, 60, 100
	Night light (neon)	5, 15
Mushroom Filament	Silica and Pearl	40, 60, 100, 150
	Pink Pearl	60, 100
	Silver Coated	60, 100, 150
Candle Filament	Clear and Silica	25, 40, 60
	Twisted, Clear and Silica	25, 40, 60
	Pink Pearl	40, 60
Shaped (Fantasie, Gaiety, Gala)	Pearl and Coloured	40, 60
Pigmy Filament	Clear, Pearl, Coloured	15, 25
Round Bulb Filament	Silica	25, 40
	Coloured	15, 35
Rough Service Filament	Pearl	25, 100
Tubular Filament	Architectural straight 3 mm, 6 mm, 12 mm	35 – 100 watts
	Curved, circles	60
	Single Cap Tubular	15, 25
	Double Cap Striplite	30, 60
Reflected Filament	Internal silvered (ISL) Spotlight or flood*	75, 100, 150
	Pressed Armoured Glass (PAR) Spotlight or flood*	100, 150, 300
	Crown Silvered (CSL)	60, 100
Tubular Fluorescent	Standard	15 to 125
	Arrow Slimline	30 to 50
	Miniature	4, 6, 8, 13
	Circular	22, 32, 40
	Reflector 2 ft, 5 ft	20 to 80

* Or Edison screw

extreme end of the spectrum and *ultra-violet* light is just beyond the visible range. Ultra-violet light makes certain chemicals glow very brightly when it shines on them and so the inside of the tube is coated with chemicals that glow or *fluoresce*. By putting different chemicals in the fluorescent coating of the tube, several of different colours can be obtained, ranging from a cold bluish-white to a warm pink-white. Deluxe warm white is the colour most used in the home but there are places where other colours can be used (see Table 3).

The advantage of the fluorescent lamp is that it gives between three and five times as much light for a given number of watts as a filament lamp; has a much longer life – between 5 000/7 500 hours – and the light it gives is almost without glare and the long tube does not cause harsh shadows. If two fluorescent lamps are arranged at right angles to each other, illumination is virtually shadow-free. Baffles and diffusers can disguise the tube and spread the light.

Fluorescents for the home come in lengths from 150 mm (4 watts) to 2,400 mm (125 watts) and in tube diameters of 15, 25 and 38 mm. Slim tubes are for concealed fittings. There are also circular tubes and some made with a reflector lining.

Every fluorescent lamp has a control gear in its circuit to help start the tube to 'glow' and to control the electric current to the correct value. As shown in *Fig. 22* there is a choke or transformer and a capacitor and a starter switch.

The general rule for choosing fluorescents is ten watts for every square metre to be lit. A single ceiling mounted tube may be sufficient for the average kitchen up to 11 square metres but two 1 500 mm tubes in a fitting – which can be mounted flush – may be needed in larger rooms.

Suitable places for fluorescents outside the kitchen are pelmets to accent the curtains (see *Fig. 23*), or mounted vertically on a wall concealed by three baffles linked together. (See plate 9)

Sometimes a fitting has a barely audible hum which can

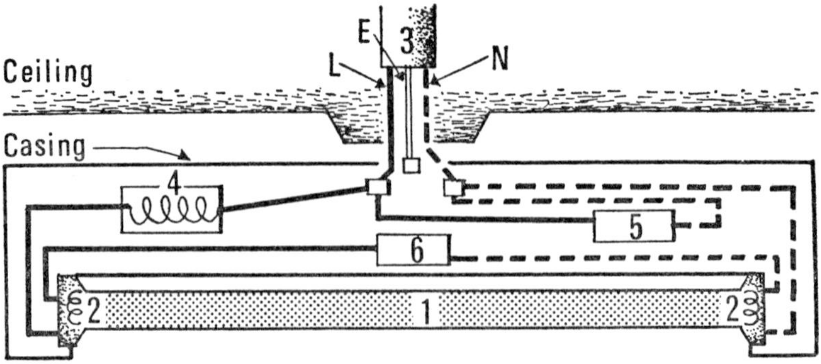

Fig. 22 How a fluorescent lamp works. The fluorescent lamp consists of a tube (1) with cathodes at each end (2). Electric current passes from the mains supply into the tube casing (3). From the live (L) terminal a supply passes through a regulator called a choke (4) while the neutral (N) passes through a capacitor (5) which works in conjunction with the choke. The current passes through the cathodes (2) and into the starter (6). The starter's job is to warm the cathodes so the current can flow between the cathodes inside the tube (1). This generates ultra-violet light which makes the inner surface of the tube glow brightly because it is coated with fluorescent salts. The earth wire is connected to the metal casing

be irritating, so always ask for a quiet fitting that will not vibrate. One cure, if you should have trouble, is to get the electrician to remove the choke from the fitting and fix away from the tube – possibly on a ceiling joist in a suitable enclosure. Quick start units can also be fitted to stop the slow flick-flick effect as they go on. There are also specially designed dimmer switches which can be used with fluorescent lamps.

Some fluorescents develop an irritating flicker at the tube ends – most noticeable where the tube is visible. Shielded cathode tubes are now sold in several lengths which reduce this and also prevent blackening at the tube ends.

If the lamp makes repeated unsuccessful efforts to start it may be the starter and not the tube and the starter is easy to replace. Don't leave a fluorescent to get old and die. To-

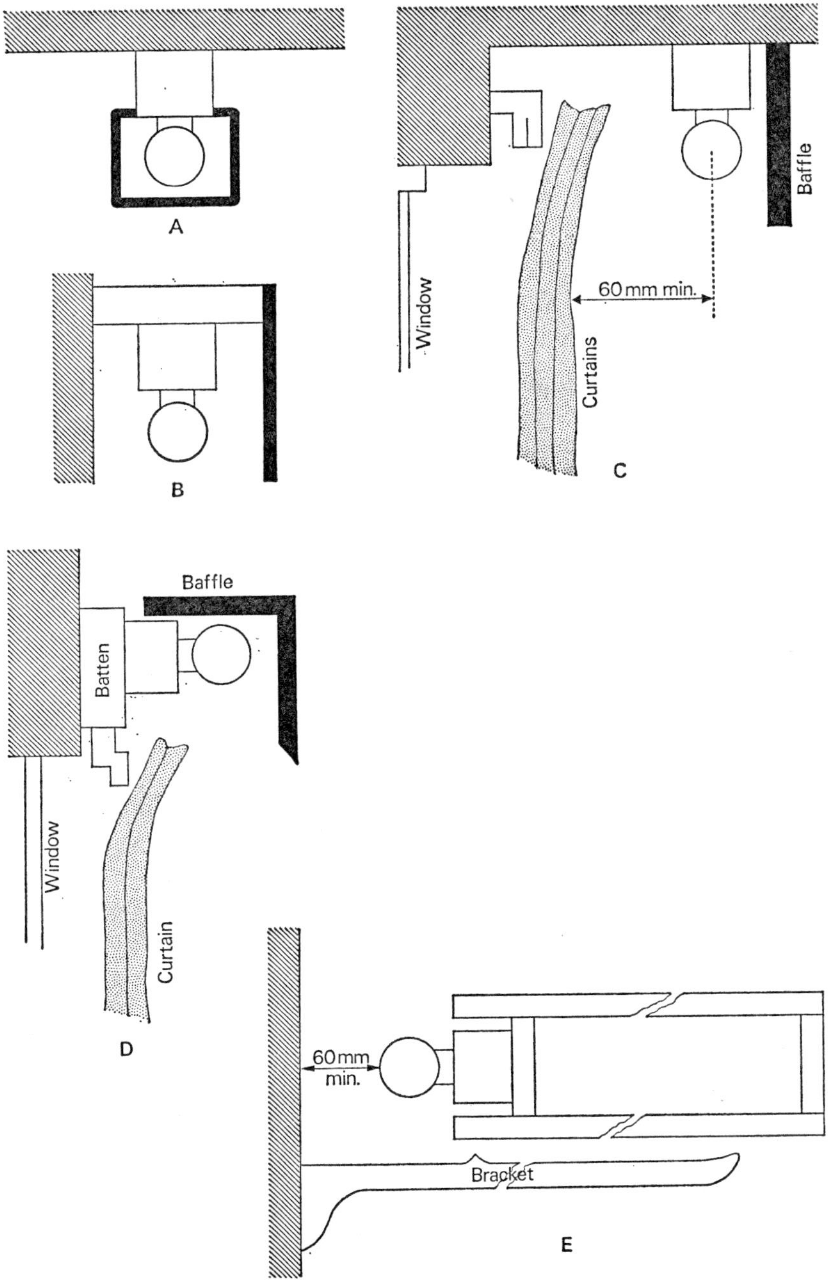

Fig. 23 Fitting fluorescent lights – some of the many uses. (A) Fluorescent light fitting with diffuser. (B) Fluorescent concealed behind baffle. (C) Fluorescent lighting a curtain. (D) Fluorescent pelmet. (E) Fluorescent shelf

wards the end of its life of 7 500 hours (six hours light a day for three years) its output will fall by 20 per cent and it is best to replace at the first sign of dulling.

The light output of every lamp depreciates as burning hours increase – normally by about 15 per cent throughout life (the exception is the tungsten halogen lamp which remains close to 100 per cent). This depreciation does not include the gradual accumulation of dirt on the lamps or on the louvres, baffles or reflectors so regular cleaning is most important.

FLUORESCENT LAMP COLOUR

There is a difference between the colour rendering and the colour appearance of discharge lamps – in particular fluorescent lamps.

Colour *rendering* is the way a lamp reveals or 'brings out' colours viewed by its light. 'White' light is made up of a mixture of the colours of the visible spectrum, and many different mixtures will be accepted by the eye as being 'white'. When colours are viewed by the light of a 'white' lamp their strength will be influenced by the proportion of their colour contained in the mixture.

Colour *appearance* is the apparent colour of the light source and also the colour of true white objects viewed by that light. One example of the difference between rendering and appearance is the warm white fluorescent lamp. This has a warm, almost pink, appearance and yet red objects viewed by its light have their colour 'flattened', that is, made less intense.

To put it simply: colour appearance is the colour of the lamp when lit; colour rendering is the ability of the light from the lamp to reveal colours in the objects illuminated.

Choice of the correct colour fluorescent lamp for a room is to some degree a subjective matter but Table 4 can be used as a guide. In general, the higher the illumination level needed the 'cooler' should be the lamp colour employed, that is, higher K. But general lighting from 'cooler' coloured lamps tends to look dismal and in the home, where the requirements are usually below 300 lux, fluorescents of below 4 000K are the best.

LAMP CAPS

Most filament lamps for the home have caps of the *Bayonet* type (BC for short) to fit into the familiar bayonet holder. The barrel of the cap has an outer diameter of 22 mm. The candle lamps and tubular filament lamps often have *Small Bayonet Caps* with a smaller diameter barrel (SBC for short), 15 mm. Almost all three pin bayonet caps are reserved for lamps designed for a special purpose such as creating the glow in fuel effect fires.

Many of the reflector lamps have an *Edison Screw Cap* (ES for short) so the lamp can screw into the specially designed holder. GLS lamps of 300 watts and above have a *Goliath Edison Screw* (GES).

Miniature Edison Screw (MES) and *Lilliput Edison Screw* (LES) are described in Chapter 6 as they are normally reserved for Christmas tree lights.

Single Centre Contact (SCC) are for filament tubes with a cap at each end. Architectural tubular lamps are fitted with special peg caps. All fluorescent tubes have bi-pin caps sized according to tube diameter.

If you need a replacement and the cap seems unusual to you, it is best to take the original lamp to the electrical shop so it can be matched. It may be difficult to obtain replacements for lamps from abroad as many foreign caps are different. It is advisable always to insist on British lamp-

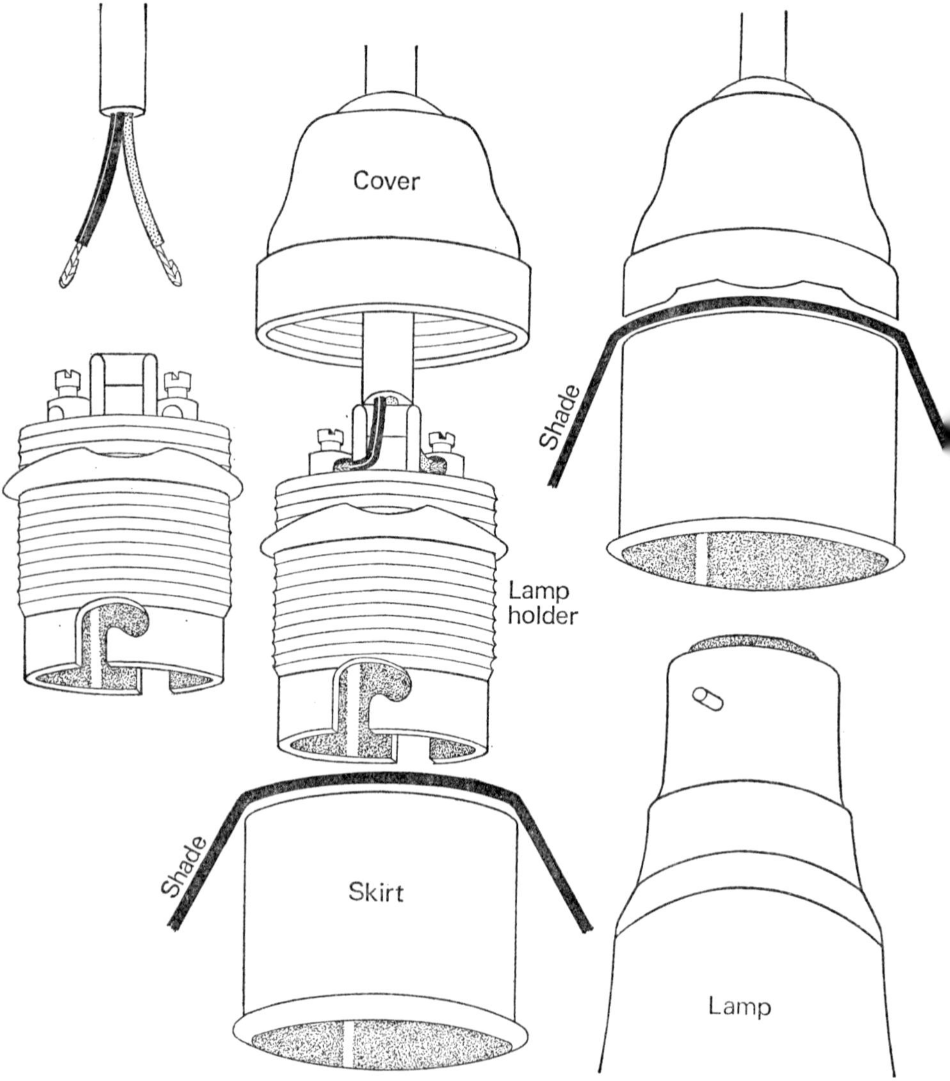

Fig. 24 Replacing a bayonet lamp holder. (*Left*) Remove the cover, unscrew the terminals, bare the flex as shown. (*Centre*) Thread the flex through the top of the holder before connecting the flex to the terminals. (*Right*) After connecting flex to the new holder, screw on top and bottom covers and shade, keeping the flex slack, then insert the lamp

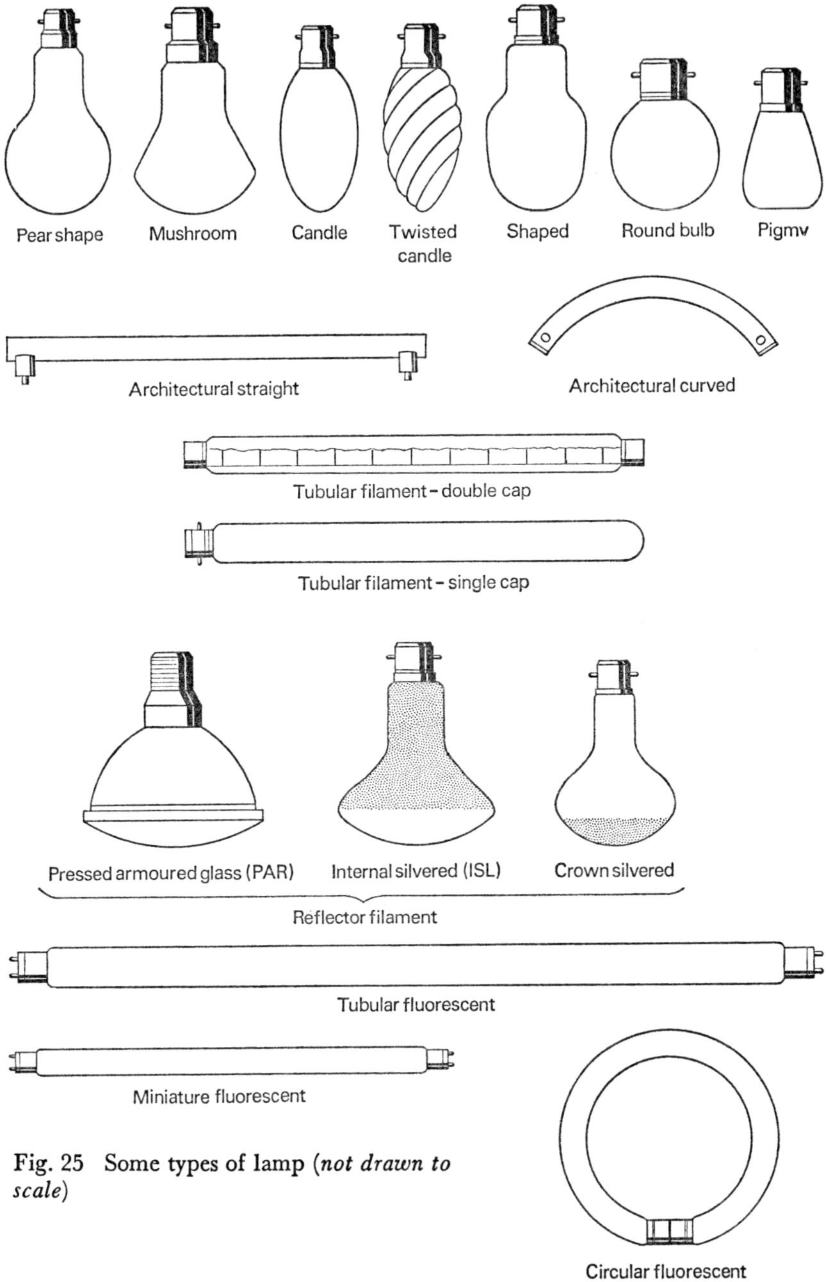

Fig. 25 Some types of lamp (*not drawn to scale*)

TABLE FOUR

GOOD COLOUR RENDERING FLUORESCENT LAMPS

Names in italics denote British Standards Colours, but the names of some tube colours are not standard between manufacturers. Some well known brands are listed and it is not intended to imply that lamps shown on one line are identical in all respects.

There are other types of tubes made for special purposes: decorative coloured lamps and ultra violet and black-light-blue used in discotheques to make fluorescents material glow.

Colour Temperature °K	Thorn (Atlas, Ekco, Mazda)	Osram	Philips	Colour Rendering
2700			Softone 27 (55)	Similar to tungsten filament and blends well with it
3000	*Deluxe Warm White* (65)	*Deluxe Warm White* (65)	*Deluxe Warm White* 32 (65)	For homes, hotels, restaurants and some shops
3600	Deluxe Natural (55)	Deluxe Natural (55)		Good on all colours, reds are emphasised Used on food displays, especially meat
3900			Colour 34 (65)	Similar to daylight with some sunlight
4000	°Kolor-rite (65)			A good general purpose colour for offices, shops and
4050	*Natural* (75)	*Natural* (75)	*Natural* 25 (75)	lighting pictures
4200			Trucolour 37 (50)	For critical colour matching especially in art galleries
5000			GraphicA 47 (60)	Complies with BS.950 Pt. 2. For critical colour matching in graphic arts; also used for general lighting
6500	*Northlight* (65)	*Colour Matching* (65)	*Colour Matching* 55 (65)	Similar to north sky daylight. Cool appearance Colour matching only to reasonable accuracy
6500	*Artificial Daylight* (40)			Complies with BS.950 Pt. 1 For critical colour matching

holders when buying a fitting – otherwise you may find yourself with a fitting and no way of getting a new lamp.

REPLACING A LAMP HOLDER

A bayonet type lamp holder will need replacing if the plastic moulding is damaged or cracked at any point, or if there is no longer sufficient tension in the contact plungers. Switch off at the main and push each plunger up and down if you wish to check the tension. Do not try to do it if the circuit is live – indeed, never try to replace a lamp with the circuit switched on. Stand on a proper step-ladder, not a chair or a newspaper spread on a corner of the grand piano.

To replace the holder, remove the lamp and shade then unscrew the cover and loosen the flex from the clamping grooves on the central pillar. Unscrew the terminals and disconnect the holder from the twin core flex. (see *Fig. 24*) It is assumed that the shade or fitting is not of metal, otherwise it will need an earth wire from the earthing terminal supplied in a modern lighting wiring circuit. Connecting an earth in the circumstances will require the services of an electrician.

Unscrew the cover from the new bayonet cap and thread it on the flex. Fit the flex ends into the terminals and tighten firmly. Fit the flex around the pillar of the holder in the grooves provided. Screw on the cover, replace shade, using the plastic ring or skirt provided. Insert the lamp then switch on to test.

If the flex to the lamp holder is worn it will need replacing at the ceiling rose. If the switch feels hot or rattles, or flashes, or does not click properly, then something is broken or worn and it should be replaced. If the lamp sparks at the holder or flickers constantly, then it is best to get a qualified electrician.

There are books which will advise on do-it-yourself home

electrics but, unless you have studied the subject carefully and undergone a course under qualified instructors, tampering with electric wiring – other than simple jobs like fitting a plug or replacing a lampholder – is not recommended.

Chapter 3: LIGHT FITTINGS

Light, as every playgoer knows, is the essence of enchantment. Lighting can be used not just to illuminate but to decorate.

Light from *above* is the easiest to arrange but it can sometimes be unflattering as the fronts of objects seem dark to the onlooker. *Frontal lighting*, on the other hand, is the most revealing since the side in view is fully illuminated, but there are difficulties in avoiding glare.

Side lighting is associated with dawn and evening when the sun is close to the horizon. Light flowing across an object brings out the texture by adding shadow and it can be dramatic and theatrical. Again, the difficulty is in arranging the light so the beam does not enter the viewer's eye.

Back lighting produces a halo effect and is usually used to separate an object from its background. *Lighting from below* will also draw objects away from their surroundings and make them appear suspended in space. It needs careful arrangement with a translucent horizontal surface and lamps concentrated beneath. Displays of glass and jade which transmit light are ideal for this treatment.

There are no fixed rules when using light to decorate but theatre lighting specialists use this guiding code: above for illumination; front for emphasis; side for modelling; back for drama; below for novelty. In the home the aim is to try to seek a combination so you can create any atmosphere by switching lights or altering the volume or angle of light.

Never buy a fitting just because it looks pretty. Never buy a fitting because it matches the carpet or the wallpaper, or the curtains. Never buy an antique wall lamp bracket just because it matches the wood of the furniture – except if you want it purely for decorative effect.

Consider the appearance of the fitting lighted and unlighted. Ask yourself: how much light does it produce? How much light does the shade absorb? Consider carefully what part of the room is to be lighted and for what purpose. Try to work to a lighting plan so that the fitting is part of a scheme. Try not to hamper yourself with the convention that confines lighting outlets to the centre of the ceiling. There they often look solitary and awkward, placed with no consideration of how the room will be used or its appearance when furnished, lighted and lived in. For a light fitting is not just a way of screening a lamp and preventing glare. It should be decorative when illuminated and provide the right kind of light, in the right place, angled in the right direction, to suit the situation.

Overhead central lighting – except in kitchens, bathrooms and on staircases – lacks interest by itself. It is often unflattering to furnishings and unsuitably placed for reading in comfort – or, in fact, for any of the usual family activities. It is much better if several ceiling points can be provided – usually without one in the centre.

A simple way of getting more service from the centre is to use it as the supply point for a section of lighting track (see page 35). The aluminium track can be mounted in or on the ceiling or on the wall and there is a variety of fittings that can be plugged in anywhere along its length.

Many of the fittings described in this chapter will not be found in the domestic catalogue or even in the domestic lighting showrooms of the big stores. There you can buy any light fitting so long as it can be suspended from the centre of the ceiling or fixed to the wall! The best light fittings for the home are in the manufacturers' commercial and architectural lighting catalogues. Outside London it is very difficult to find anywhere to view the fittings so you should write for catalogues – which are all well produced and well illustrated.

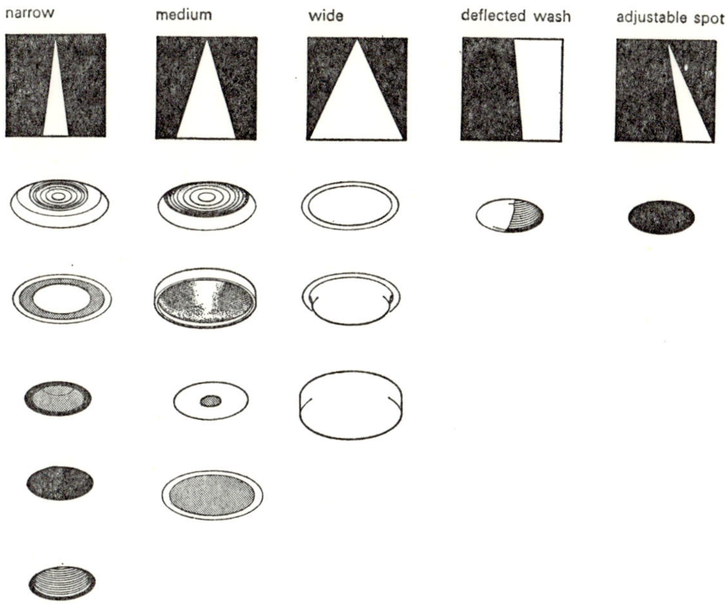

Fig. 26 A selection of different types of cylinder downlighters (Rotaflex)

Do not be put off by the technical jargon. If you have read this book carefully you will understand all the essential parts anyway.

Fittings should be used only with the lamp sizes recommended by the makers. Larger lamps may not only make them too bright for comfort and cause unpleasant glare, but may harm the fitting.

CYLINDER DOWNLIGHTS

These are sold under a variety of names, such as highlighters, and are designed to concentrate light from the ceiling. Most use reflector lamps, some GLS, and all are classified by beam spread (*Fig. 26*): wide and medium for general lighting; medium and narrow for accent lighting; deflected wash pattern for vertical surface lighting, or adjustable spot lighting.

The fittings can be fully recessed into the ceiling, set flush with the surface, or semi-recessed so that part of the cylinder protrudes, or they can be mounted on the surface. There is a variety of finishes and the majority hide the lamp from view behind a lens. Some of the lamp holders are at the top of the fitting, some have the holder at the side – used where space between floors is limited.

Installation is not difficult – a circular hole is cut into the ceiling and the fitting held in place by special clips. Replacing the lamp is quite simple but get the electrician to show you how to do it at the time of installation.

Up to four cylinder downlighters have been put into a ceiling and the wiring connected to the former centre ceiling connection, but this does depend on the loading of the lamps and the loading on the circuit.

PENDANT FITTINGS

The pendant fitting is the most popular form of lighting for the home. It is also usually the worst designed: or perhaps it would be more fair to say that it is not very often used in the correct situation. (See *Fig. 27*)

Multi-lamp pendant chandeliers give much less light than their size and expense would suggest. Hung too near the ceiling, many of them make semi-circular pools on its surface. But they do give a good diffused side light and, if properly supplemented with portable lamps, they can be effective. Most of the smaller ones – two and three-lamp size – are best regarded as decorative fittings and additional forms of general illumination must be provided.

Some of the many pendant lights are shown in *Fig. 27* with an indication of the type of light they give and a suggestion on where they might be used. The best situations for these fittings are staircases and passages where they can add to the decorative effect. For pendants are hang-down lights that are always obtrusive and, in situations where they have to compete with the furnishings, they often come off second best.

16 Cheerful room designed for family living

17 Same room with party time lighting. (Pictures are mounted on a baffle board concealing fluorescent lamps)

18 (*Above*) Modern living room showing use of electrified track and cylinder downlighters (Concord Rotaflex). 19 (*Below*) Dining/bedsitter with a rise-and-fall pendant over dining table and ceiling recessed spots for lighting ornaments. Centre ceiling fitting gives good background light (Thorn)

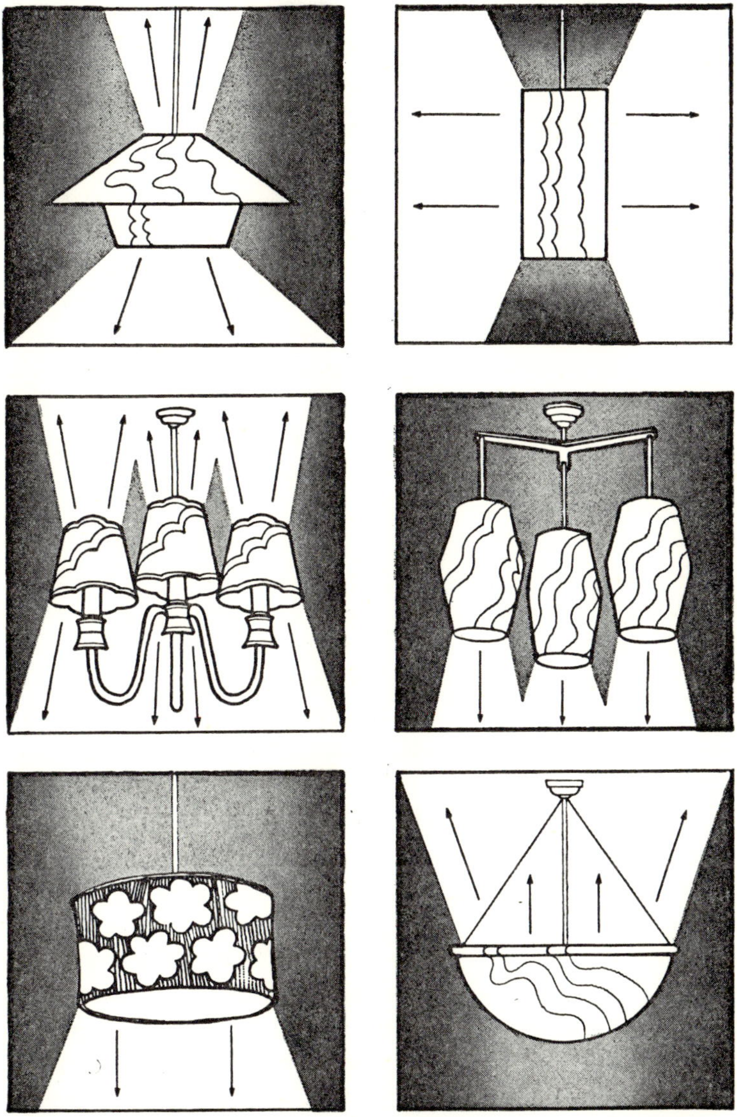

Fig. 27 Six types of light fittings. (*Top left*) Low-level fitting with diffusing shade lined white: down light for dining table. (*Top right*) Translucent shade with opaque screens for sideways light: tall rooms, passages. (*Centre left*) Multilamp chandelier supplies decorative background lighting: living room. (*Centre right*) Open-ended centre fitting directs most light into a pool: dining room. (*Bottom left*) One lamp holder in traditional shade: bedroom. (*Bottom right*) Diffused light bounced from white ceiling: unsatisfactory

E

One place where they are useful is over a dining table. They are often best with a counterbalance (rise-and-fall) fitting. The pendant light may have its own spring suspension arrangement so that height can be varied at a touch but it is more usual for the rise and fall mechanism to be available on its own – either in the ceiling rose or a ball on the cable suspension. The light can be pulled down or taken up by hand to the right height but – being used at a low level – the fitting must diffuse the light without casting glare into the diners' eyes. Direct light from the open base will give sparkle to silver and glass, but there must not be too much or you will have glare and unpleasant shadows on your guests' faces.

Passages, staircases and halls need cylindrical shades. Much of the light should be given off horizontally to avoid the pool effect produced when most of the light is directed upwards and downwards.

Many pendant lamps have heavy shades. The weight of the suspension, holder and lamp alone is frequently 0.3kg ($\frac{3}{4}$lb) so it must be securely fixed to the ceiling. Heat resistant flexible cords must be used and the suspended weight on the cord must not exceed certain limits depending on the flexible cord used (IEE Wiring Regulations). For example, the commonly used 0.5 mm flexible cord is limited to 1.3 kg (3lb) fitting weight.

It is also possible to loop pendant fittings from a centre ceiling point. Lengths of flex are taken in a long loop to an insulated hook over the point where the fitting is wanted. Although this is a simple answer to centre ceiling point problems, the disadvantage is that it is never very sightly and the flex collects dirt which is difficult to remove.

SURFACE FITTINGS

Pendant fittings very often are made with matching wall and close ceiling mounted versions. Many of the close ceiling fittings can also be mounted horizontally but, before you buy,

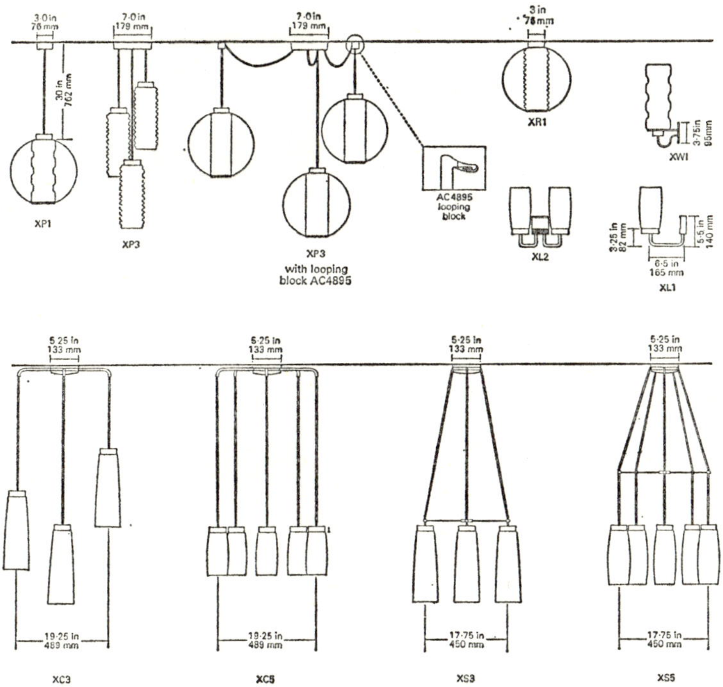

Fig. 28 Samples of pendant and wall fittings from The Thorn Lighting Catalogue

look first at how they spread their light, as patterns that look nice on the ceiling can look ugly on a wall. No wall fitting should ever be placed where it can cast glare into the eyes. There is no recommended standard height for wall fittings – it depends on the fitting – but it is usual to place the fitting so the lamp is slightly above eye level.

On a staircase with limited headroom or a hall which traps the wind when the door is open, a close surface fitting may be far better and safer than a pendant.

Wall lights can produce a more pleasant effect than ceiling lights but, as the light intensity must be kept low to avoid glare directly into the eyes, with its consequent discomfort, the result is that illumination begins to fall off

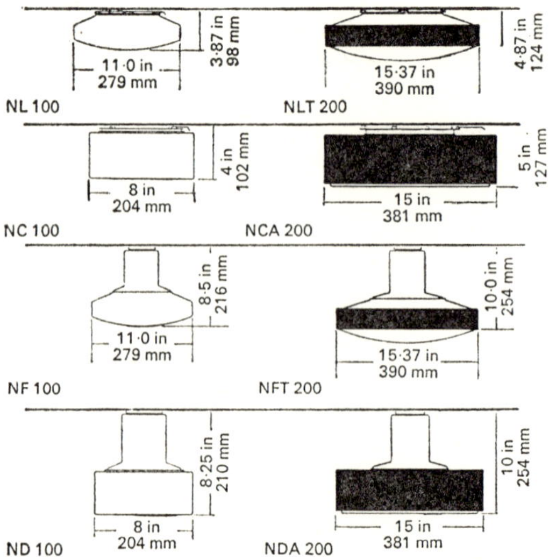

Fig. 29 Ceiling surface fittings from The Thorn Lighting Catalogue

about two feet away from the wall. Rooms tend to look smaller when lighted mainly in this way.

Wall lighting is most successful in large rooms or where a directional fitting, such as a spotlight, is hung on the wall – designed so that its angle can be changed to fit the mood of the room – on to a desk or sideboard, on to the ceiling for softer lighting at entertaining time. It is best if conventional small lamp wall lights are used for decorative effect rather than for bright illumination.

LIGHT BRICKS

These are rectangular, square and circular opal glass lighting units – most of them designed for surface or recess mounting indoors or in protected positions outdoors. Most are about 100 mm deep. Weatherproof versions can be used for house identification as numbers and letters can be applied permanently to the glass.

ACCENT FITTINGS

This class of fitting – sold under a variety of names – is growing in popularity because of its versatility. In the living area of the home the need is for flexible lighting. Reflector lamps mounted in swivel fittings can place a pool of light anywhere the width and size of beam varies only with the fitting. For complete flexibility they should be mounted into an electrified track but they can just as easily be mounted on walls, ceilings and on or under shelves.

Fig. 30 Reflector lamp mounted in swivel fitting (Rotaflex)

Wide-beam or flood lamps are best for short throws or low floodlighting levels. Narrow-beam or spot lamps are best suited to long throws and higher levels of spotlighting or display lighting. Pin-spot lamps are for long throw, high-intensity lighting of small objects like pieces of sculpture or floral decorations.

Colour filters can be fitted to most of the accent fittings. They cut down the *amount* of light but enhance mood and create drama.

A clean, light coloured surface may reflect three-quarters of the light that reaches it; a dark or dirty surface may reflect a quarter or less. So, for equal display brightness, the dark surface will usually require about three times as much illumination as the lighter surface. It is important to remember this when choosing the lamps and fittings. Table 5 is a guide to the performance of the reflector lamps.

Note, too, that a light coloured object against a dark background will appear to be brighter, with less light than a dark object against a light background.

TABLE FIVE

REFLECTOR LAMP PERFORMANCE GUIDE

Lighting Intensities at 3 metres from light source*

Lamp	*Diameter*	*Wattage*	*Type*	***Light output*** *(at centre of beam)*	*Beam size*
PAR	170 mm	300W	Wide Flood	970 lux	1.68 m circle
PAR	170 mm	300W	Medium Flood	1940 lux	0.86 x 2.11 m
PAR	170 mm	300W	Narrow Flood	3350 lux	0.63 mm x 1.35 mm
PAR	120 mm	150W	Flood	320 lux	0.53 mm x 0.74 m
PAR	120 mm	150W	Spot	810 lux	2.21 m. circle
PAR	120 mm	100W	Flood	190 lux	0.81 m circle
PAR	120 mm	100W	Spot	430 lux	2.21 m circle
ISL	95 mm	100W	Spot	80 lux	0.81 m circle
ISL	126 mm	150W	Spot	430 lux	2.21 m circle
ISL	126 mm	150W	Flood	110 lux	1.70 m circle
GSL	60 mm	100W	Cream Silvered	480 lux	2.85 m circle
GSL	80 mm	150W	Clear	300 lux	0.97 m circle

* For intensities at 1.5m multiply by 4; intensities at 2m multiply by 2.

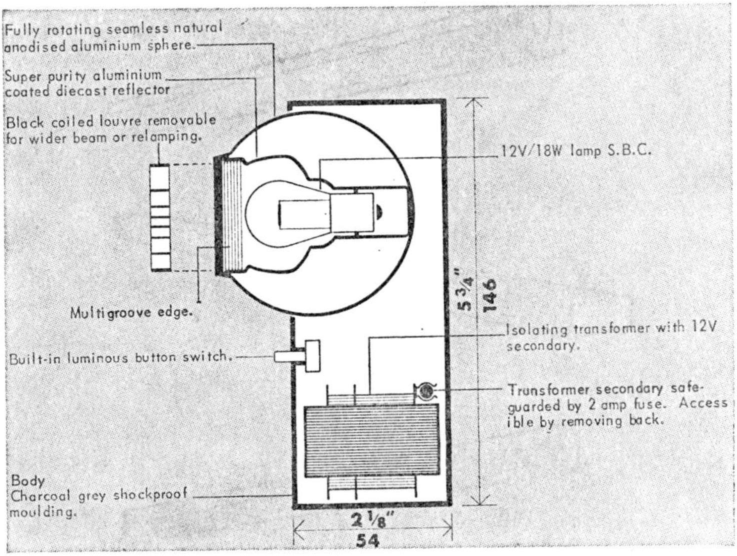

Fig. 31 Accent fitting: narrow beam, miniature, high intensity fitting (see Fig. 30)

Among a variety of fittings, some of them illustrated, is the *miniature spotlight* and the *framing spotlight*. The framing spotlight, made by Rotaflex, give a high intensity shaped beam for illuminating pictures, sculptures, displays or special architectural features. Concealed shutters can be adjusted to produce a variety of quadrilateral and triangular shapes. The spotlight uses a low voltage tungsten-halogen lamp which has a life of 2,000 hours.

FLUORESCENT FITTINGS

For a long time the fluorescent lamp was confined to the kitchen. Use elsewhere in the home was declined because the lamps were unwieldy and difficult to disguise. Prejudice was fostered by the choice of the wrong colour tube and the clinical, commercial design of the fittings which did not seem to blend easily with home decorations and, since the majority of people buy fittings on their shop appearance, they

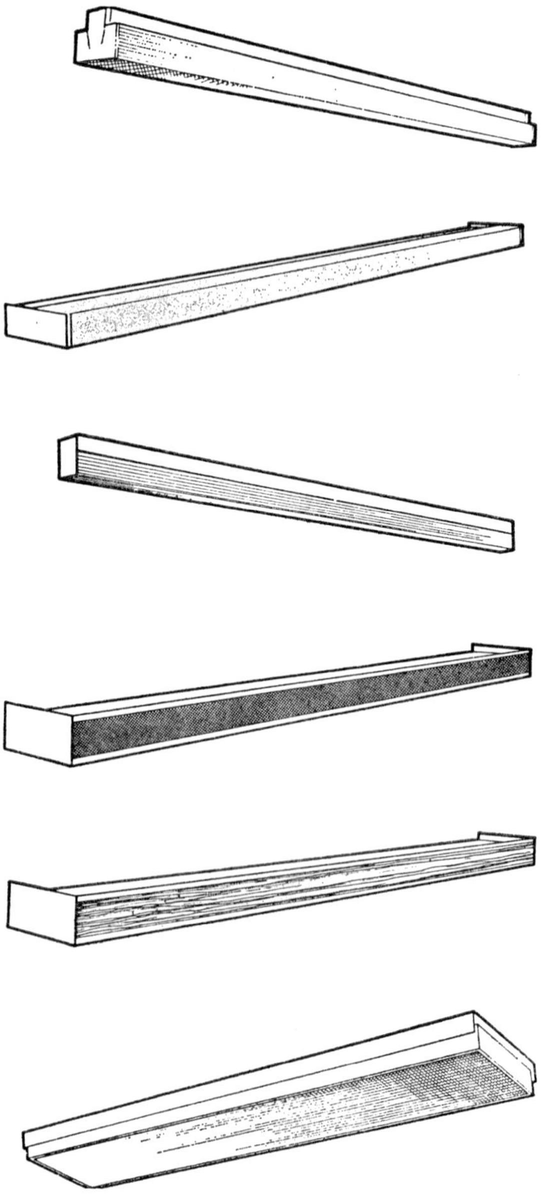

Fig. 32 Fluorescent diffusers – a few of the many different types

were difficult to display. For the way to approach the use of fluorescents in the home is to say to yourself: here is an inexpensive, very efficient light source. How can I make use of it in my home?

Start by looking at the manufacturers' catalogues; look particularly at the single lamp diffusers, pelmets and baffles – some of which just clip on to the tube. Or you can conceal a lamp up in a curtain pelmet or behind a structural beam inside luminous panels. The lamps can be fitted inside a louvered cornice, inside a false ceiling or set vertically on a wall disguised by baffles – front and both sides. Some of the many ways they can be used are illustrated.

If a fluorescent lamp is mounted less than 228 mm below the ceiling it will produce a bright band of light. This can be reduced by adding a diffusing plastic top or front, or the lamp can be concealed inside the pelmet – a minimum of 60 mm from the curtains – then there is no spill on the ceiling.

A fluorescent can often be mounted between the joists in the ceiling behind a diffuser. There is usually sufficient air

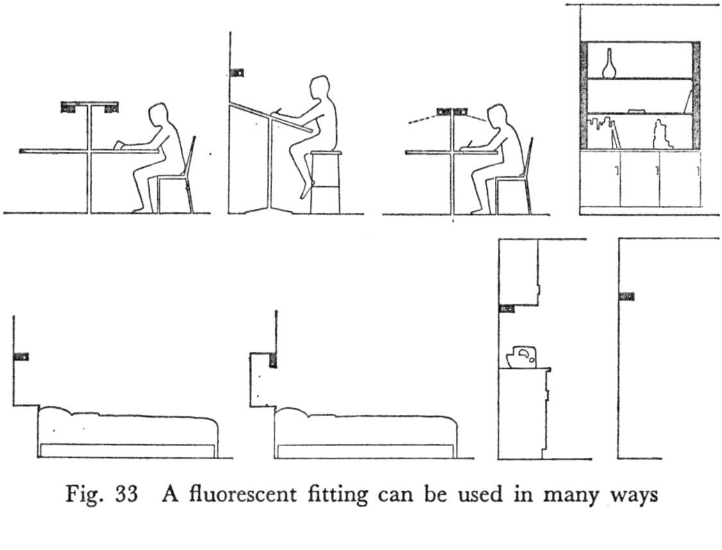

Fig. 33 A fluorescent fitting can be used in many ways

circulation between the joists to prevent overheating. It is often best – although it is not essential – to let 50 per cent of the diameter of the fluorescent tube project below the ceiling surface to ensure a reasonable spread of the light output. The tube can be totally encased by a suitable plastic louvre or opaque cover – of which there is now a very wide range.

Circular fluorescent lamps can be fitted inside bowl or circular diffusers. They look very nice in bedrooms and study/bedrooms for general lighting.

Some of the slimmest lamps with a variety of coloured baffles are made by Atlas.

PORTABLE LAMPS

Portable lamps are generally picked for their design, their shape and colour, rather than for their merits as a source of light. This is natural as, for the greater part of the day, they must please on their ornamental merits alone.

Shades should blend both in colour and brightness rather than contrast sharply with their backgrounds. As a general rule shades should be lined with white and have reasonable sized openings both top and bottom. Plain shades are usually the most practical. Nymphs or dryads look better chasing each other round the base rather than on the shade!

In small rooms bulky shades on portable lamps can hinder rather than help the decorations. An alternative is the 'tree' standard where one, two or sometimes three small fittings are mounted on a pole and can be angled to place light where it is wanted. Some will take either reflector lamps (spotlights) or ordinary lamps. Be careful with spotlights in portable lamps as they can cause glare unless carefully angled.

It is important to test the shade in the shop with a lamp, for the colour may change dramatically with the warm light. Although shades are best if they have a white or light coloured interior or lining to reflect the light rather than hold it back, the amount of luminance – how bright it looks – depends on material rather than colour. Plastics are often better than fabrics or parchment – the latter materials are better than metal or tinted glass. You can now buy washable shades in nylon that can be taken off their wire frame and put into the washing machine.

Success as a source of light depends on shape as well as the strength of the lamp. A double ended tube with the lamp set in the centre will create two pools of light – one up and one down – probably very decorative but not practical as a light source.

At least one lamp in the room – table or standard – should be chosen for chairside use for reading and have a shade that will spread the light with the base wider than the top. The lamp socket must not be in the centre of the shade but in the lowest position.

For a standard lamp the minimum height from the floor to the top of the shade should not be less than one and a half metres, and, for a table lamp, the minimum height from the table to the bottom of the shade should not be less than 300 millimetres, otherwise there is poor distribution of light. The standard lamp shade should be capable of screening a 150-watt lamp and the table lamp shade at least a 100-watt lamp.

Decorative lamps with enveloping shades can only be used as luminous features. Metal or pottery lamps pierced with holes are very attractive but are not suitable for reading. Even for general use they illuminate too small an area and can cause glare unless the brightness of the immediate surroundings has been brought up to a comparable level by light from other fittings. However, they can can be very useful in a child's room where they will give a cheerful nursery glow.

There are other times when a low level of light is needed and there are now table lamps incorporating a dimming switch in the base so that the luminance can be adjusted. Individual dimmer fittings to put between lamp and socket cost about £5 for the 300-watt version and can be adjusted by turning a knob on the side.

Never plug other appliances – such as an electric iron – into the lamp socket or into a portable lamp. It is dangerous as the lamp is not earthed and also it can be easily pulled over and broken. Avoid, too, whenever possible, trailing flexes across the floor. Ideally, no lamp should be more than a few feet from a socket outlet but where it becomes unavoidable to extend a lamp flex, buy a properly designed connection or rewire the lamp. Do not resort to tape or exposed twisted wires.

CUPBOARD AND SHELF FITTINGS

Walk-in cupboards, larders and built-in wardrobes should all have their own light. Some have a switch that is worked by opening the door which is not difficult to fit.

Alternatively, choose a fitting with a pull-cord or press-button switch.

Shelves can be lighted from above using a miniature fluorescent lamp fitting or tubular filament lamps.

PICTURE FITTINGS

Specially designed fittings are made for lighting paintings, using a tubular filament lamp; effective, but the fitting is obtrusive and it often means an exposed flex. A profile spot to give a patch of light the same size and shape as the canvas is the ultimate refinement (see illustration facing page 48) but a spotlamp that will light the picture and its immediate surrounding is a very satisfactory compromise. Best results are with pictures that are framed without glass.

READING LIGHTS

Miniature reading lights that can be adjusted to any angle to spread light over a book are useful for desks and bedheads. They contain their own low voltage transformers and switches but they cover only a small area and light output is rather limited.

Chapter 4: PLANNING YOUR LIGHTING

Begin by deciding what is to be lit. Where do people sit in the living room? Where is the TV set? Where is the desk? Where does Mother do the sewing? Remember, too, that every area where activity takes place will need to be lit and the lighting solution should be flexible. Lighting for one type of task may be entirely inappropriate for another.

Having decided what the lighting has to do, it is much easier to decide where the fittings have to be placed, and it will help if a plan of the area can be sketched roughly to scale using squared paper. In areas like the living room, which is the family centre, allow for a number of points of comparatively low wattage which will provide a much more acceptable solution than a single large wattage source which can cause glare and is inflexible.

Decide what light distribution will be required from the fitting and do not choose fittings which all give the same type of light distribution. This can result in monotony and, in the case of concentrated direct downward light, unpleasant shadow effects. The light distribution of fittings is divided into upward, direct and indirect. The most useful general lighting fittings for use singly are those of a 'semi direct' distribution which, whilst giving most of their light downwards, permit some upward illumination.

The floor plans on pages 78–81 show how typical homes have been planned with a good standard of lighting.

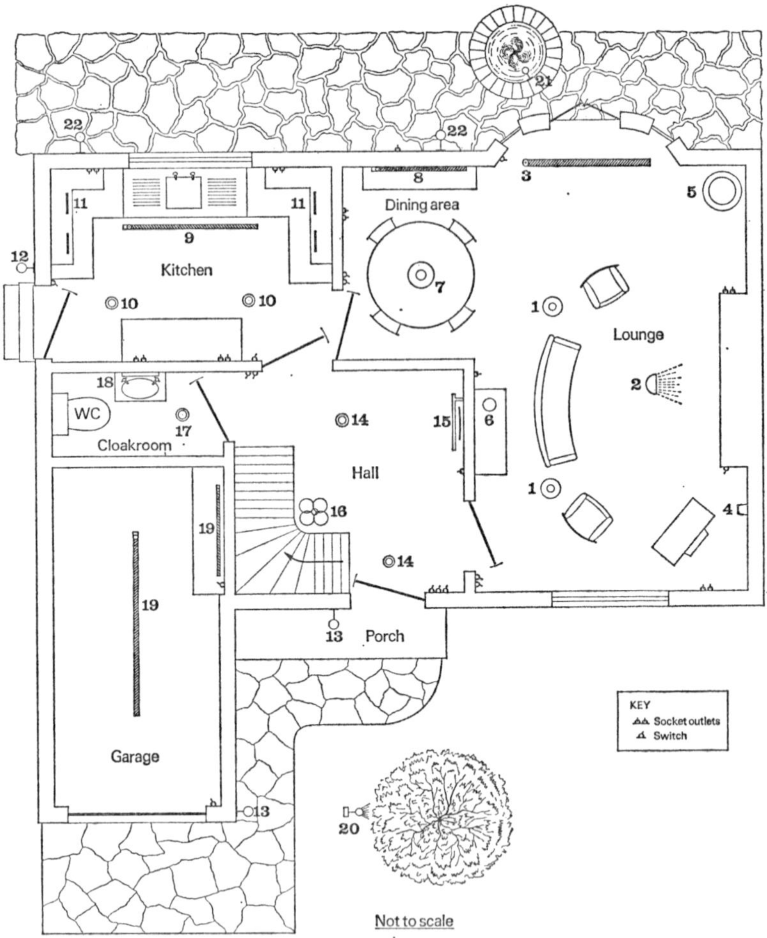

Fig. 34 Planning your lighting. (A) (1) Downlights recessed into ceiling space give reading light in sitting area. (2) Recessed angled downlight lights wall and painting over fireplace. (3) Fluorescent lamp recessed between ceiling joists behind a diffuser panel lights bay. (4) Fluorescent mounted vertically on the wall (see picture facing page 32). (5) Standard lamp. (6) Table lamp. (7) Adjustable pendant fitting over dining table. (8) Fluorescent mounted horizontally on wall with baffle lights sideboard. (9) Ceiling mounted fluorescent lights sink. (10) Cylindrical downlight recessed into ceiling. (11) Mini-fluorescents under wall cabinets light worktops. (12) Outdoor wall fitting lights kitchen steps. (13) Matching outdoor fittings light path and porch. (14) Recessed cylinder downlights. (15) Picture light. (16) Cascade of pendant lamps light hall, landing and stairs (two way switched). (17) 60W ceiling fitting. (18) Strip light over mirror. (19) 6ft fluorescent for ceiling; 3ft fluorescent over workbench. (20) Garden spot-light angled into tree branches. (21) Garden fountain illuminated by pool light. (22) Pair of patio lights – wall mounted

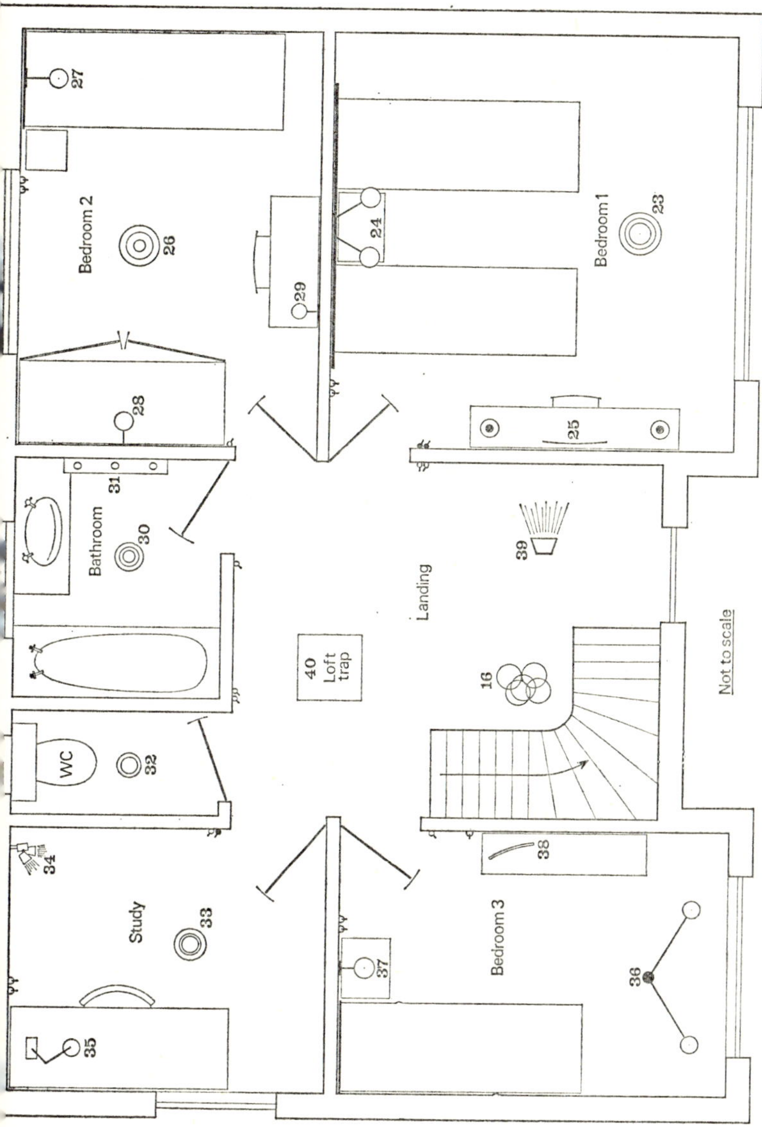

(B) (23) Flush-to-ceiling fitting with two 100W lamps. (24) Twin bed lamps. (25) Pair of pendant lamps suspended each side of dressing-table mirror. (26) Ceiling surface light. (27) Bedlight fitting mounted to wall. (28) Built-in cupboard light switches on as doors are opened. (29) Wall-mounted light for desk. (30) Flush-to-ceiling recessed fitting. (31) Mirror light with shaver socket outlet. (32) Flush-to-ceiling mounted fitting. (33) Recessed cylinder downlight. (34) 4ft section of electrified track with three spotlight fittings. (35) Table lamp. (36) Pair of pendants looped from ceiling point and hung either side of window. (37) Bedside light. (38) Illuminated plug-in make-up mirror. (39) Ceiling-mounted spotlight lights painting or sculpture. (40) Loft light

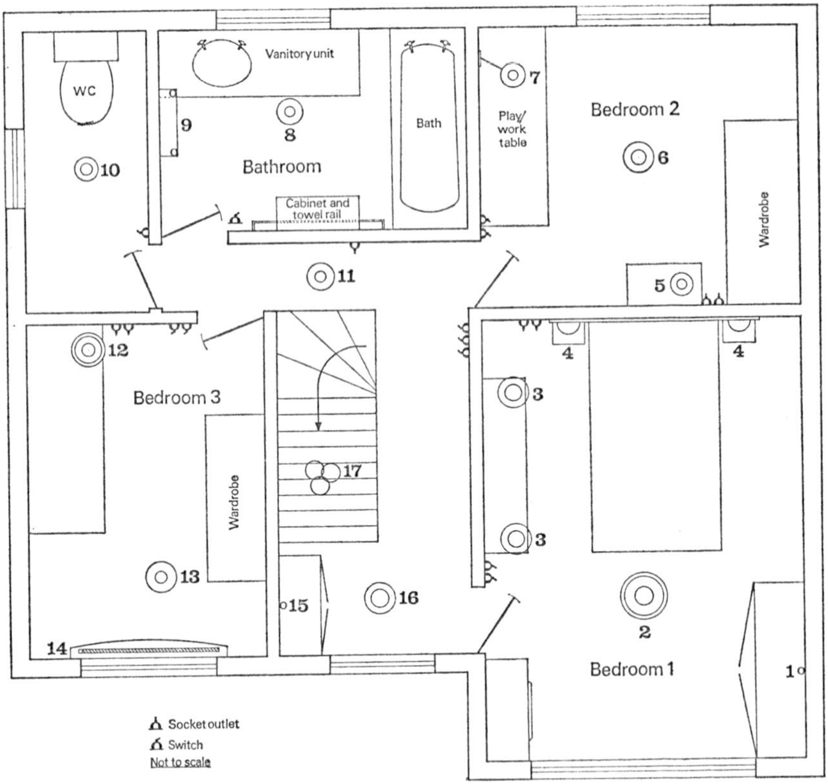

(C) (1) Built-in wardrobe with light (door switched). (2) Flush ceiling fitting. (3) Twin pendant each side of dressing table. (4) Bedhead light. (5) Bedside light. (6) Flush ceiling fitting. (7) Wall light on moveable arm. (8) Ceiling recessed light (pull cord switch). (9) Mirror with twin lights and shaver socket. (10) Ceiling recessed light. (11) Ceiling downlight. (12) Pendant light. (13) Flush to ceiling fitting with matching shade. (14) Fluorescent in pelmet. (15) Cupboard with built-in light. (16) Cylinder downlight. (17) Pendant cascade

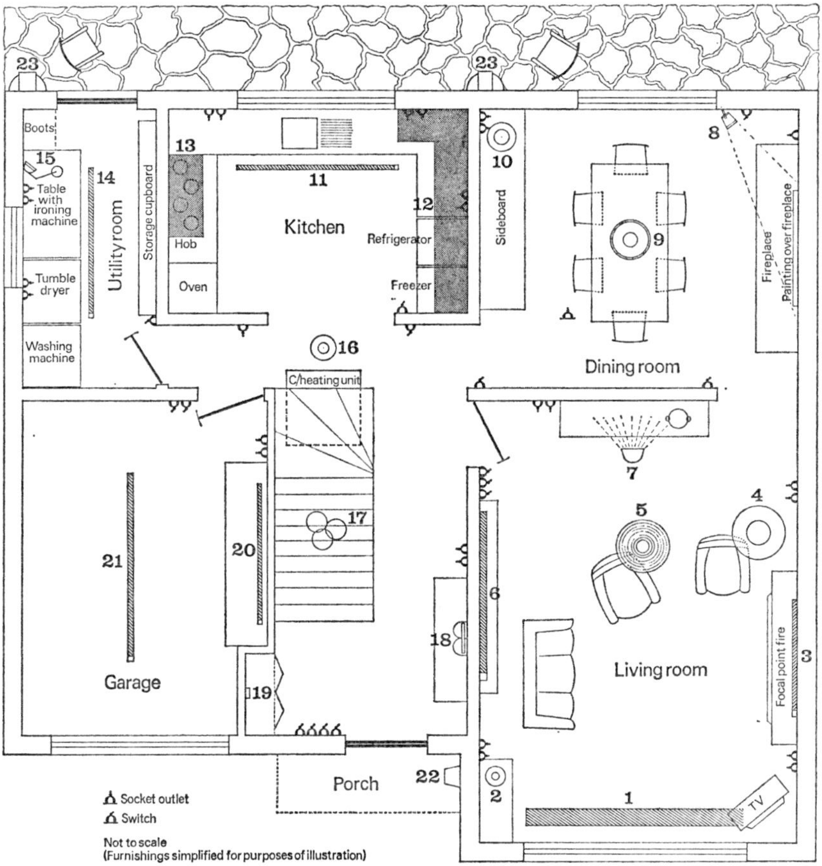

(D) (1) Fluorescent fitting between ceiling joists. (2) Table lamp. (3) Picture fitting or mini-fluorescent lighting painting over fireplace. (4) Floor standard. (5) Flush to ceiling fitting (dimmer controlled). (6) Fluorescents in wall unit. (7) Angled downlight. (8) Angled framing spotlight. (9) Rise and fall pendant. (10) Table lamp. (11) Fluorescent. (12) Slimline fluorescent under cabinet. (13) Hob hood with light. (14) Ceiling fluorescent. (15) Angle lamp. (16) Ceiling fitting. (17) Pendant cascade. (18) Wall light over table. (19) Coat cupboard with light (door switched). (20) Workbench with fluorescent. (21) Fluorescent. (22) Porch light. (23) Weatherproof lights

LIVING ROOM

General light can be provided in various ways. One of the best is from recessed downlighters in the ceiling to which local light is added, then spotlamps could be used to highlight the curtains. Remember that deep colours will absorb light; pale or bright colours reflect light. The advantage of using spots is that they are adjustable. If, for instance, there is a wall unit which is more interesting visually, the spotlights could be related to that instead of the curtains which could, alternatively, be lit by a fluorescent fitted above or inside the pelmet. It is wonderful the difference that lighting one wall can make to the general appearance of the room.

A point worth remembering; the further the lamp is away from the surface it has to light, the wider the spread of light across the surface will be, and the less will uneven plastering be noticeable.

Wall lights are not always successful. A modern alternative to wall brackets is to use a fluorescent tube screened by a baffle, and fixed horizontally about half way between floor ceiling. This would unify a group of pictures arranged above and give a working light on to a settee or radiogram beneath, as well as providing a general background light over the rest of the room. Several patterns of this type of fitting are now sold, including one of aluminium which has no paint finish to yellow or chip and one which can have a fabric or wallpaper inset to match (Rotaflex).

Display cabinets can have small lamps incorporated, either tubular striplite or 15 mm diameter fluorescent. Pictures and sculpture or an attractive vase should be highlighted.

Take care where you put the TV set in the room. When viewing television no direct reflections of light sources should be visible in the screen.

The aim is to have 25 watts of lighting for every square metre even though, on average, only about 10 watts per square metre will be in use at one time.

Reading light is needed at each armchair and writing

table, usually provided by portable lamps which must be of a suitable height (see page 75). To avoid long trailing flexes to the centre of the room cylinder downlights or flush fittings may be let into the ceiling instead. Reading lamps should not be less than 100W and floor lamps not less than 150W.

Dimmer control of ceiling lights would give the greatest degree of flexibility – high for family living, down through several degrees according to the atmosphere required.

There are two main types of dimmers – *resistive* and *solid state*; a third type using thermionic valves is usually only employed on very large loads such as theatre lighting. Resistive dimming uses a resistor which dissipates some heat, and this can sometimes be a problem. Solid state dimming produces very little extra heat, but there may be a slight flickering effect from the lamps when they are dimmed well down.

Obvious uses for the dimmer include TV viewing (especially colour) listening to hi-fi, intimate dinner parties, nursery or sick rooms, all-night lighting on landing or stairs, hall porches and showing slides or home movies.

Cost of a dimmer is about £5. Make sure that the unit you buy – and there is a wide selection on the market – meets the GPO Specification BS 800 for interference suppression.

Fluorescent lamps can also be dimmed – although this requires additional ballast units and resistors or a specially designed dimmer.

There is no noticeable colour change when a fluorescent is dimmed but filament lamps become increasingly red as they are dimmed. The manufacturers usually specify the load which a dimmer will handle safely and without overheating this may be expressed in watts or amperes.

Check that your dimmer has short circuit protection. A violent voltage change can sometimes damage thyristors and replacement is expensive – it usually means buying a new switch. Although this is a rare occurrence, if your home is subject to regular fluctuations in voltage, it may be better not to install dimmers.

If in any doubt as to the correct use of dimmers seek advice from the maker, a reputable electrical contractor or your Electricity Board.

High ceilings in older houses give opportunities for lighting either by creating a false ceiling that conceals light fittings, or painting the existing ceiling a dark colour and suspending a light coloured platform over the central area which is used to contain the cylinder downlighters or surface mounted fittings. Another method is to use a darkish colour on ceiling and walls down to the desired level then use a bright wall covering to the floor. Suspend pendant fittings – the type that shed their light downwards – level with the colour break on the walls.

DINING ROOM

The need here is to provide a soft general light and a light over the dining table. A *pendant fitting* that directs most of the light downwards will serve the dining table, or possibly a chandelier over the table with dimmable candle lamps. The fitting must be positioned so that it does not interfere with the view across the table or cause glare when looked at from a sitting position.

A pendant on a rise and fall fitting is often the best method (see page 66).

Cylinder downlighter fittings will provide general subdued lighting and if their wall switch is replaced by a dimmer switch a dramatic setting is achieved at little extra cost. Otherwise background light can be provided by a portable lamp on a sideboard (100W), a floor lamp in a corner (150W), by a wall bracket (60–100W each) or by lights mounted in a section of lighting track.

By throwing light on to a wall, curtains or paintings, the interior can be visually expanded. This form of lighting is called 'wallwashing' and can be achieved with reflector lamps fitted into the ceiling or on the walls.

A dining room is also the room where many of us display our ornaments and family heirlooms. Slim fluorescent fittings mounted on the back edge of a shelf on a wall unit will create the effect of 'floating in space'. The tube should be mounted 60 mm from the wall (see page 53).

The same idea can be used to make a luminous shelf over a storage heater or central heating radiator but here, because of the heat, architectual filament lamps should be used. The wiring, too, must be heat resistant.

HALL AND STAIRS

Falling is a major hazard in the home. The elderly, who must contend with less efficient eyesight, have the most frequent accidents. Many could be prevented with good lighting.

Stairs and passageways are danger areas, but because they tend to be treated as 'dead' space, they get rather less attention for lighting and decoration than they deserve. Yet it is when a person comes out of a brightly-lit room into a dark hall that accidents often happen.

To walk upstairs safely there must be clear distinction between stair treads and stair risers. There should be a lamp at the top of the stairs which will brighten the tread and leave the risers in shadow. A second lamp at the foot of the stairs or in the hall is needed to give general illumination so the shadows are not too dense. The lamps must be shaded so there are no patches of glare on the eyes. It is best if both lamps are on the same wiring circuit and they should be two-way switched – both at the top and bottom of the stairs.

Where a staircase changes direction or there is a half landing, the top light should be position where it can give good illumination to both sets of treads. Sometimes this means placing a light in an inaccessible spot so, to avoid dangerous acrobatics when changing the lamp, it may be a good idea to fit a rise-and-fall unit.

If there is sufficient height available, hang a cascade of three or five lights from a ceiling point on the first floor (60W or 40W each). This will give plenty of light where it is most wanted. If the landing changes direction or there are cupboards, an additional ceiling fitting will be needed.

The porch and the entrance hall should be lighted to give visitors a warm welcome. The lighting level can be less than in the living area as eyes need time to adjust from the outside darkness. Pick warm colours. You will find that yellow or orange tinted shades in the hall give the best effect.

The lamp to light the front doorway should be totally enclosed and weatherproof unless it is inside a porch and protected from wind and wet. Even there it must be protected from damp. Several fittings are now made to take the house number or even the name – which helps identification in the dark. (Plate 14)

Safe lighting for paths, fuel stores and outside steps need not be costly. Enclosed fittings can cost less than £2.

THE KITCHEN

A kitchen is full of shiny, reflecting surfaces. Pots and pans, dishes and glass are all light reflectors; cupboards, the cooker and other appliances interrupt the light from a centre lamp. So, wherever you stand in the kitchen, it is difficult to avoid shadow or glare. To provide good, overall general lighting the ceiling-mounted fluorescent is perfect as it gives a 'line of light' spread over a large area.

Two – or even four – tubes can be mounted side by side in a fitting. There are also circular tubes for circular fittings but the straight version is normally best for the kitchen as it should be lined up above the sink unit – slightly over the front edge and parallel to it to give glare-free lighting.

You can also use fluorescents for supplementary lighting of work surfaces set against the walls. The tubes can be mounted under an overhanging cupboard and hidden by a narrow wooden pelmet or a plastic baffle. Light here is important as you will probably be shielding the main light source from illuminating what you are doing. In a small kitchen with reflecting walls and work surfaces that bounce back the light, this may not matter: too much local light may cause glare. But with wood panelling or dark vinyl papers, and where there are overhanging cupboards, some local light is essential.

Walk-in cupboards and larders need their own light and the easiest way of switching on and off is by fitting a door latch switch. The light cannot then be left on accidentally.

One other way of providing glareless lighting is to conceal either spotlights or tubes on top of cupboards that are not full height and so bounce light off the ceiling. Wiring can be hidden in the cupboard to supply under-cupboard lights for worktops. Or you can dispense with fluorescents altogether and just use a group of cylinder downlighters – or a combination of both as in the kitchens illustrated in plates 21 and 22.

Many cookers have their own hob light and many ovens and larger refrigerators are also self-illuminating – this sets an example for improving lighting in specific areas.

The kitchens illustrated apply the good lighting rule – good general plus local illumination – but the variations are endless.

Another method – achieved with the help of the architects and builders – is completely concealed lighting. The entire ceiling is illuminated with the light source concealed behind diffusing panels. An advantage of this type of ceiling is that it can be used to conceal a venting system to remove steam and cooking smells.

BEDROOM AND NURSERY LIGHTING

A centre ceiling light switched only from the bedroom door is not just unnecessary but hazardous – as every nocturnal toe stubber will testify. For bedrooms look best when indirectly lighted – even when the room is used as a refuge from television or as a place for hobbies and homework.

Lighting should be concentrated around two main areas – the bed and the dressing table. Built-in or walk-in wardrobes should also have some form of lighting; so should a full-length mirror.

All principal lamps or ceiling lighting should be capable of being switched from both bed and doorway. Two-way switching as described in Chapter 1, is both simple and not expensive if planned while the house is being built or rewired.

The common fault with bedside lighting (apart from not having any) is to place the light source too low. Small, squat lamps on a bedside table give a pleasant diffused light but are no use for reading in bed. Unless you prefer to read with the aid of a torch under the bed clothes, a bedside light should be at eye level or slightly above. This can cause discomfort when lying down as the light is then thrown directly into the eyes, so either an angled lamp or an enclosed shade should be chosen. Stability is most important. Don't pick a lamp that can be knocked off easily. Clean, dry sand in the base of pottery lamps will help, but make sure it does not leak out of the flex inlet (wax or modelling clay will prevent this). Best answer is to have the lamp fixed to the bedhead or set on the wall above. The lighting for a double bed should be switched separately from each side and the recommended minimum lamp is 60 watts.

Twin beds usually go for one lamp on a table between them. This is fine if there is agreement on when reading is to stop and sleeping begin. But a small fluorescent in a bafflle or an enclosed lamp set on the wall above each bed is more harmonious in every sense. The flex can be hidden behind the bedhead but should be pinned to avoid accidents when moving the beds.

Bedhead lighting is often needed during daylight hours. If the bed faces the window, looking at books against strong light can cause strain – particularly to an invalid – and a

20 Lighting a stair well: Five light pendant (Osram) hang as a cascade on black cable from five-way ceiling plate (Falks). Two of the glasses are clear, two are smoke, one pale amber. Each takes 40W lamp and light on each stair tread is 60 lux

21 (*Above*) For a high ceiling, and where concentrated illumination is needed, downlights can be recessed partially or completely into the ceiling. In Diana Rigg's kitchen three Rotaflex Multigrooves provide general lighting in the kitchen area and Atlas fluorescent lamps are hidden behind a plain wooden baffle to light the worktop. 22 (*Below*) Completely shadowless lighting in the kitchen can be achieved by mounting fluorescents at right angles along the ceiling above the edge of the work surfaces. Plastic 'egg box' louvres in wooden frame disguise the lamps and can be easily removed for cleaning. Dining area is lit by partially recessed cylinder downlights

23 Lighting a kitchen worktop with fluorescent concealed below cupboard

24 (*Above*) Modern bedroom: twin pendants (one shown) light either side of the dressing table and wall lights – independently switched – provide bedside light (Thorn). 25 (*Below*) Bedlight to serve twin beds. Fluorescent disguised behind pierced metal baffle, mounted on hardwood block extended to house two-way architrave switches at both ends

strong light on the printed page is needed to balance the strong frontal sunshine.

Of course, the difficulty is that bedside lighting has to compete with the electric blanket for use of the socket outlet. Adaptors get knocked when the bed is moved so the safest solution is to replace a single socket with a double socket mounted flush to the wall.

Getting rid of the centre ceiling pendant is usually simple. If there is easy access to the loft above the lighting position it can be moved or even multiplied using extra cable and junction boxes. Moving the switch position may be more difficult but it costs nothing to get advice.

Having the ceiling pendant moved so it hangs between you and your reflection over the dressing table makes making-up easier. If the dressing table has to go into the window space then extra lighting will be needed to offset the strong light from the window. Ideally, the mirror should be lit both top and sides but twin pendants about one metre apart will do very well and the mirror will also reflect light back into the room. Remember, it is the face and not the mirror which has to be lit so a single pendant must never be immediately over one's head but at least half a metre up from the tip of the mirror.

Fluorescent tubes fitted into the curtain pelmet can be wired into the ceiling system and switched from the door. This will provide the diffused lighting and make the room look comfortable. As a fluorescent needs to be set a little away from the curtains, a false pelmet may be needed.

Every built-in wardrobe fitting should have its own lighting and here the wiring is easy to conceal.

In the nursery a nightlight is needed – both for the child's own sense of security and for parents' inspection without having to use the main light. Bedside lights can be too easily over-turned so a wall-light with a low wattage (15 watt) lamp is one simple answer. Running costs add only twopence a week to the electricity bill and later – as the child gets older – the lamp can be replaced with a 40 or 60 watt lamp.

Another way of lighting the nursery is to install an *electro-luminescent* nightlight. This is a picture which glows. It is mains operated and provides sufficient illumination to see one's way round the room. Power consumption is so low that it only costs a few pennies a year to run.

This electroluminescent material in the form of tape is likely to be more widely used in the home for lighting in the future. Light is produced by exciting a luminescent material laid in a thin layer between flexible electrically conductive plates; one of them transparent to allow the light to emerge. The amount of light produced is not great but sufficient to mark hazards and fix round switches or keyholes so that they can be easily located in the dark.

THE BATHROOM

The bathroom is the nearest equivalent to a private, personal beauty-parlour most women possess. Yet builders of bathrooms usually only provide a single light in the centre of the ceiling so that very little of the face can be seen in a mirror.

The opal glass bowl set in the centre of the ceiling effectively holds back something like 40 per cent of the light output. Many of them are badly designed and are a better cemetery for flies and other 'wee beasties' than a source of illumination. There are some good fittings but in large bathrooms two will be needed or otherwise one fitting capable of taking a lamp of high wattage – 150 watts or two of 60 watts.

The regulations covering wiring in bathrooms are strict and builders choose the easy and least expensive way out – a single lighting point set safely out of reach in the centre of the ceiling! Very difficult to supplement once the plastering and tiling is completed. The only kind of socket outlet permitted by the regulations is one made specially for electric shavers which contains a device isolating the user from the mains. It must not be used as a plug for portable lighting.

Like electric heaters and towel-rails, light fittings must be permanently fixed and connected directly to the circuit wiring. Pendant lamps are allowed but are not usually suitable because of steam and condensation.

The regulations are designed to avoid the possibility of anyone touching any metal part of any electrical fitting while standing or lying in the bath, or when using the washbasin. Electricity and water do not mix. Should – by some rare mischance – the metal become 'live' and anyone were to touch it while immersed in the bath or touching the taps, the resulting shock could be fatal.

Electrical accidents in bathrooms are rare. One serious one was caused by a record player propped across the corner of the bath and connected by lengths of flex to a bedroom lighting point. It fell in. No regulations can protect against this kind of thing – only repeated warnings.

The easiest way to improve the lighting in most bathrooms is to improve the output from the centre light. Choose a fitting of plastic and glass that is both corrosion and dust proof, suitable for a 100 watt lamp – minimum.

A fluorescent tube will give the most light for the lowest wattage but choose one made to withstand steam – rusted contacts can mean early tube failure. A fluorescent can also be recessed into the ceiling space, except where there are solid floors above.

Any bathroom light fitting should be installed by a qualified electrician. There must be no exposed flex and the fitting – if it has any metal parts – should be connected to an effective earth. The switch must not be on the light fitting itself unless it is of the pull-cord kind and independent switches should be either pull-to-the-ceiling or wired to a switch outside. The lamp fitting should be steam-proof.

If you decide on fluorescent lighting try a 'de luxe natural' or a 'natural' colour tube rather than a 'de luxe warm white'. 'Natural' is nearer to the colour of sunshine daylight and is better for making-up. If you have areas of white tile, however, de luxe warm white is advisable. Electricians – if you ask them nicely, will try out both colour tubes for you so you can make a decision.

If you are building or replanning your bathroom then run cable into the wall beside the bathroom mirror. As well as serving a shaver socket, the cable can be concealed behind the mirror to serve a light. Two lights set each side of the mirror and placed so that they stretch from the level of the chin to a little above the head is the best way. A tube or a lamp set above the mirror is not, on its own, as helpful in making-up as side lighting – as any actress can confirm.

Mirrors that have the silvering removed at the edge to allow light from a built-in lamp to come through tend to throw light directly into the eyes – causing glare. The silvering also gets ragged at the edges after a time.

Heat and lights units are a good idea. The light source is either a filament lamp or a fluorescent tube and the heater is usually the radiant kind. Although the fitting fits to the ceiling in the place normally taken by the centre light, it should not – because of its loading – be connected to the lighting circuit. The unit should be connected – via a spur – to the ring circuit or have its own circuit with an earth.

Often ways can be found to get better lighting in the bathroom without spoiling tiling or the ceiling. A floor to ceiling mirror mounted on the tiles is one way, with wiring concealed behind.

A fluorescent tube fitted against the cornice can throw light down one wall of a larger bathroom and can be operated from the main switch. Or recessed cylinder downlights can be used to supplement the centre light, angled on the basin.

A downstairs cloakroom can be lighted with a simple enclosed ceiling fitting or a glass wall fitting as illustrated facing page 49. There should also be a mirror light for the benefit of visitors anxious to appear at their best. Various mirror fittings with an integral light are available or a small fluorescent with a baffle could be installed.

LOFTS

A light should be installed in every loft, switched from below. The switch should have a neon indication as a reminder so the light is not left on accidentally. If the loft is to be turned into an extra room then it should be lit in the same way as a bedroom or a playroom.

GARAGE

A standard industrial fitting – preferably fluorescent (65W) – is quite adequate, although local lighting over a bench will be needed if home or car repairs are done here rather than in a workshop.

Chapter 5: OUT OF DOORS

There are practical as well as decorative reasons for garden lighting. Lighting steps, paths and doors prevents accidents and helps you to recognise callers quickly. A house which has good exterior lighting is less attractive to the housebreaker.

But the justification for garden lighting is that it can reveal an unsuspected beauty in the familiar; a favourite conifer lit with a single spotlight and viewed after dark through a picture window adds a pleasant sense of spaciousness; an underwater light adds magic to an ornamental pond. After all, why should we not make the garden come alive at night? So few weekend gardeners are home early enough on a weekday evening during a large part of the year to enjoy the results of their labours that the black pool behind the house seems hardly worth the effort.

The equipment to light the garden is not expensive and it is not difficult to install unless your ambition tempts you to create a son-et-lumière. However, you will not be very satisfied with the results if you string up the lights willy-nilly or place a floodlight just any old how, hoping for the best. If you have noticed the subtle way in which the lamps in local parks have been sited so that the most attractive features of a subject are highlighted, then it is hardly necessary to stress the need for careful thought and experiment before finally deciding upon the colour, position and angle of the light.

The ideal position is a matter of moving a lamp from one site to another until the most attractive effect is achieved. A leafless deciduous tree, for instance, can be brought to life by directing the beam of light up through the branches.

Shield the lamp so that the light source is not apparent and the overall effect is a soft, diffused light.

A greenhouse or glass-sided conservatory or shed can be used to create a striking effect after dark by installing fluorescent tubes. Line up pots and plants so that they are silhouetted against the glass. The fluorescent light source can also be used to stimulate plant growth.

Small town gardens, patios and even 'area' take on a completely different look after dark if they are ingeniously but not over lit. If you have a single ornament, statue, tub or plant container it can be picked out by one or two floodlights, casting fascinating silhouettes against a wall or paving and making the garden look much larger. But make sure that you do not angle lights so that they shine directly up into neighbours' windows.

The walled garden can also be attractively illuminated by mounting the specially designed outdoor fittings at intervals on the brickwork. There are several floodlight fittings for outdoors now – many of them using the tungsten halogen lamps.

The main connection to outdoors should always be done by a qualified electrician. The socket outlet outside the house must be of the all-insulated type and it is best to have the waterproof rubber pattern. Connection to an outdoor socket must be made with armoured cable, or with the Pyrotenax type, mineral-insulated and copper-sheathed, buried at least half a metre deep in the ground.

Any connections in the cables must be electrically safe. Never, in any circumstances, use 'taped' joints. Special waterproof rubber connectors are available for this purpose in 2 amp, 5 amp and 13 amp sizes. Make sure that the female component is on the 'live' side so that the connector with the pins (male) is on the lamp side.

The flexible cable used must be tough enough to stand exposure to the weather as well as a certain amount of ill treatment. When selecting the sites for the floodlights, remember to check the route of the connecting cable to avoid possible damage by the lawnmower, edge trimmers or any other garden tools.

In a garden it is what is lighted that matters – not the lights themselves. They should be concealed in a flower border, hidden in branches or behind shrubs.

If you are lighting a path, the lamps can be covered with reflectors which not only hide them but direct the light on to the paving. Where concealment is difficult – the edge of a lawn, a treeless area or a patio, – mount the lights so that they are above the normal line of vision. And for exposed positions choose fittings that look good by day.

Concentrate on lighting trees and shrubs rather than flower borders. Conifers look best if a concealed spotlight is aimed so that the light touches the edge of the branches. Trees with a high canopy, like elms, should be spotlighted upwards into the leaves from almost immediately below. If a tree has an interesting trunk, aim a light so the beam just touches the main limbs.

Unless the flowers are tall, direct lighting of borders can be rather disappointing. Low mushroom fittings give the best effect. Be careful with coloured lights. White is safest. It produces the most natural effect and brings out colours best. Yellow makes green foliage look sickly and dead. Red turns foliage brown. Green will bring out the colour of foliage and grass but it is an unnatural green. Blue gives a somewhat mysterious, unearthly look to things, but the effect on blue spruces and plants with pale green foliage is interesting. If you do use colour then experiment first with transparent plastic before buying colour lamps and always use pale hues.

The best place for colour is where it is accenting garden features like statues or ornaments. A blue lamp concealed at the end of the garden will draw away insects from the terrace where you are sitting. There, use a pale yellow lamp which will attract less flying 'beasties' than a white lamp.

Most garden lighting equipment operates at mains voltage

26 The Surrey home of Mr Gordon Biggs has a beautiful garden that deserves to be seen at night as well. It is superbly lit by cleverly placed exterior background and spot lighting (Homes and Gardens)

27 Figures and sculpture are more 'natural looking' when lit from above, as seen in sunlight in the daytime. To achieve this, equipment can be positioned in trees or on buildings near to the object. For a more dramatic effect, the statue can be lit from beneath to create modelling and deep shadow (Concord Rotaflex)

28 Exterior lighting should be treated, not merely to eliminate darkness, but to create space perspective and atmosphere – avoiding flat 'even' lighting which defeats all these (Concord Rotaflex)

so precautions must be taken to ensure safety. The Moonglow garden lighting made by Elsworthy is safe as it operates from a 12-volt supply. Transformer and length of cable cost about £7 and pair of spotlamps about £6. The lamps look rather like headlamps and can be screwed to a wall bracket or spiked stick and angled to any position. Almost any number of lamps can be connected to the one transformer and – because of the low voltage – this is one system that can be installed by the amateur.

There are lighting sets that work from mains voltage but are double insulated so no special earth precautions are normally necessary. Where metal parts are exposed a proper earth must be provided and maintained. A device known as an earth leakage circuit breaker is often used in place of the conventional earth as it 'trips' – that is, switches off – the supply instantly if there is a fault.

A festoon lighting set is the outdoor equivalent of Christmas tree lighting – the important difference being that it has been made to withstand the weather. One popular set has eight metres of cable with 12 lamp holders designed so that the metal cap of the lamp is covered, giving complete protection. You can get extension cables and packs of various coloured 15 or 25 watt lamps and 'flashers' which wink the lights on and off. These are extra to the basic set which costs about £3. The lamps cost about £1.75.

WATER LIGHTING

There are now floodlight fittings, insulated and completely watertight so they can be submerged in a pool – a very dramatic form of garden lighting, particularly where there is an overhanging tree like a willow. As an alternative you can light the shrubs round the edge of the pool and let the water act as a black mirror. This is sometimes effective with a swimming pool which can reflect a lighted tree in its surface. A swimming pool should also have lighting so it can be used safely on summer nights. A pair of outdoor post top lanterns, mounted 3 m high at the ends or opposite corners of the pool is one recommended method. Coloured underwater lights add enchantment but must be fixed so they cannot entangle swimmers.

If you have a small stream, conceal lights among the plants at the edge rather than among the stones. Or you can spotlight it from a tree with an outdoor spotlamp. If you want to light a waterfall aim so the water appears white against a shadow background.

If you have a fountain – easy and inexpensive to install at the same time as the garden lighting – place a waterproof fitting with 75 or 150W reflector lamp in the pool at the base of the jet and aim it straight up. If it is a spray fountain aim upwards from the centre of the fountain.

Cost of garden lighting is low. But if it is something you plan to do on a large scale, then ask your Electricity Board about tariffs. It may pay you to take advantage of the White Meter tariff which charges electricity used at night at a low price.

Chapter 6: PUTTING ON A LIGHT SHOW

Fireworks with music, son-et-lumière and now the world of pop have successfully linked sound with light to create a spectacle that is emotionally satisfying to the eye as well as to the ear. Discotheques vie with one another in producing what the informed call audio-visual stimuli and what the participants know as 'being turned on'.

Some of the effects created can be overpowering and are more likely to provoke a headache than enhance a musical performance. The use of stroboscopic lights (either of visible or 'black' ultra violet light) in some of the discos are of doubtful value and, balanced against the amusing flicker effect they give with a moving subject, are the very real dangers of triggering epileptic fits in anyone prone to them. The reasons are complicated but probably involve the co-phasing of light and music frequencies with natural brain frequencies.

For the most part the colour images are enjoyable, suit the music and enhance the mood. Some of these light/sound systems have therapeutic possibilities, helping relaxation as well as improving enjoyment of music, and there is some evidence for suggesting it as a cure for insomnia.

The simplest form of lighting for visual effect is one which is unrelated to the music. It provides ever-changing patterns of colour on a back projection screen or convenient wall or ceiling.

This is easy to create and there are low voltage reflector lamps fitted with colour wheels that will form constantly changing patterns. They can also be used with dimmers. You can buy small reflector lamps linked to a colour wheel.

Lumitron make a low voltage unit complete with wheel for about £7. It can be table or wall mounted to direct light up or down.

Then there is the cholesteric light show. At Stanford University Microwave Laboratory in California, investigations are being carried on into the colour changing properties of so called cholesteric liquid crystals. These have possible uses in many ways because they are sensitive to extremely small temperature variations. They can be changed from red to blue by a 0.2°C rise in temperature. Sound signals are induced to produce localised heating effects which, in turn, cause colour changes which can be projected on a screen or a wall.

The best method so far is one where a series of lights are fed with the audio signal from a tape recorder or a record player – or even the TV set or radio if they have a suitable output outlet. The light intensity varies and changes with the signal level.

In Britain, a commercial unit is sold by a firm using the likely name of Chromosound and, for the benefit of those with some audio knowledge, this is how it works:

The unit has three very high gain amplifiers, each responding to a particular bandwidth of the auditory spectrum. Thus the bass, middle and treble notes each have their own amplification system – the output of which is sufficient to provide 1,000 watts of lighting.

The unit also has a frequency analysing circuit which splits the sound input into the three component frequency bandwidths so that, if a piano chord is played, the unit will analyse it into a so much bass, so much middle and so much treble frequency notes. The 'impulses' from each bandwidth are passed on to the amplification circuits.

The output from each circuit is not merely dependent on the presence or absence of a particular frequency in the sound input but is dependent on the *amount* of that particular frequency present. The unit, therefore, gives an output that is dependent on the relative intensities of the three auditory bandwidths used in the analysis stage.

If you follow this so far, then the outputs from the unit are connected to a number of lamps with filters of the three primary colours. The light projected on to a white surface by these lamps will constantly change colour and saturation in strict relationship to the sound fed into the unit.

The light output from the unit can be utilised in a number of ways. It can be projected on to a screen (white or coloured), it can be back-projected through almost any translucent material. The maximum power output from the unit (3,000 watts) is such that it can be used to floodlight a building and produce a son-et-lumière effect or, in the home, a number of reflector lamps can be connected to provide low level, background lighting effects.

Chromosound also make a perspex cube that is wired and fitted with colour bulbs and filters so that when it is connected to the control unit the whole cube changes colour according to the sound input.

If you have some knowledge of hi-fi and home electrics it is not too difficult to design your own unit. A mono amplifier (valve – not transistor) connected to the tape recorder output socket of a stereo amplifier will produce sufficient power to drive a string of small bulbs (series-connected). The bulbs can be coloured with a spirit based marker fluid. The mono amplifier is connected in parallel with the channel of the stereo amplifier producing a sound which varies in intensity and is directly dependent on the sound heard. Instead of a series of bulbs, a single bulb can be used linked to a moving colour wheel.

If you are not a hi-fi expert then you can try some simple effects using metal foil and a 12 volt reflector lamp linked to a flasher unit. If the transformer is built in to the lamp fitting the flasher may be difficult. Cut the tinfoil into different sized shapes and tie them together with strong cotton, leaving about 15 cm of thread between each. Hang strings of these from the ceiling – best near a window where the draught will move them – and shine the spotlight on them.

If there is a photographer in the family, a slide projector can be used as a light source. Abstract patterned transparencies can be fed through slowly or, if the projector will take the old-fashioned glass slides, then you can try for a psychaedelic bubble effect. Paint one side of a clear piece of slide glass with a glass paint from an art shop, such as Vitrina, then put a few drops of chemical alcohol (costs pennies from a chemist) on to it. Press another glass on the top and paint the surface of that with a different colour, then add another drop of alcohol. Press on a third glass so that you get a three-tier glass sandwich with two painted fillings in between, and bind the edges with a strong tape. Put this into a projector and, when the alcohol gets hot, it will make dancing colour bubbles and slow moving colour streams which can be projected on walls or ceiling.

There are plenty of other ideas. Try aiming a reflector lamp so that it bounces light down or through a glass vase full of glass marbles. Any colour transparent surface – provided it will withstand heat from the lamp – can be used to create effect and – if you can make it move – then you have a light show.

Ultra violet and black-light-blue fluorescents can be used temporarily in place of the normal tubes for a party. Anything that fluoresces naturally nearby (like nylon) or has been treated with a fluorescent paint or dye will become easily luminous. Ordinary lights will reduce the effect so they are best turned off. Use the ultra-violet light just for short periods with suitable music.

ART LIGHTING

Quite apart from special occasions like parties, some of the lighting in the room should always be used purely for decoration. Think of using lighting as an art form : chunky pottery shapes giving light from the inside can completely alter the appearance of furniture and decoration. These are to be found in arts, crafts and pottery shops rather than lighting showrooms.

It is even possible to find coffee tables that have been turned into lamps. White, rectangular illuminated tables costing from about £13 are now sold in many furniture stores. On the continent translucent marble or coloured glass lit from below by fluorescent tubes has been tried for dining table tops but this is perhaps something for an *avant garde* restaurant rather than the home.

Many modern artists use lighting and mini-motor-power in their work. In New York there is an art studio where everything on sale moves, whistles, flickers or glows!

CHRISTMAS LIGHTING

Christmas is the ideal time to try experiments with lighting. Try a reflector lamp set low down in a window and direct the beam on to a tree in the garden, dressed with coloured balls and stars. Light a Christmas wreath on the door by exchanging the porch light for a spotlight fitting – or simply exchange the lamp for one with a warm yellow or red tint.

A traditional part of Christmas is the tree decorated and lit with a string of coloured lights. Set in the window or in a porch it is the essence of Christmas and a delight to children. Christmas lights are in strings of 12 or 20 lamps in various colours and shapes and can be plugged into a bayonet lampholder or a socket outlet. These lamps have MES (Miniature Edison Screw) lamp caps – the same as a torch bulb. Larger strings of up to 70 bulbs use the tiny Lilliput (LES) screw.

There is a British Standard for Christmas tree lights so, if possible, buy a set which has the British Standard kite mark or uses the words: 'made to the British Standard'. Always buy from a reputable source for, remember, a dry tree is a fire hazard and – while lights from well-known makers are made as safe as possible – you cannot be too careful.

Remember, too, there is no rule that says you must use the lights on a Christmas tree – they can run round a picture rail, the edge of an alcove or along a pelmet. Use transparent adhesive tape rather than drawing pins for fixing.

Always test the lights before dressing the tree. Most are wired in series (*Fig. 35*) which means that if one lamp fails all the others will go out.

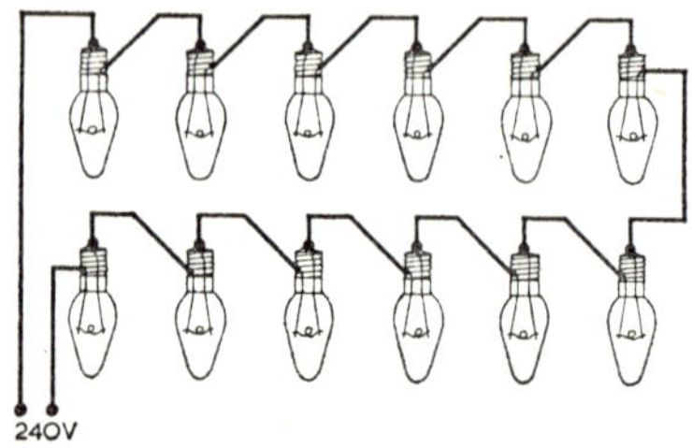

Fig. 35 Twelve lamp festoon wired in series

Lay them on the floor, free of all tangles, and make sure every bulb in the festoon is intact and making proper contact in its lampholder. Lamps which have worked loose in their lampholders present no problem. But, if one has failed, it must be found and replaced. But which is the dud? Finding out is a slow business. Remove each bulb one at a time and screw in a spare bulb and test. But if more than one bulb has failed this method will not locate the duds so it is easier to test them individually, using a battery connected to a wired lampholder.

The battery need not be of the same voltage as that of the lamp rating. A 6 volt lantern battery, or even a $4\frac{1}{2}$ volt flashlamp battery, will test most bulbs though the low voltage will show only a dull glow of the filament, not always discernible. Some MES bulbs and some in other shapes can be tested in a flashlamp or torch having an MES lampholder.

LES bulbs can be tested direct from a battery by using a

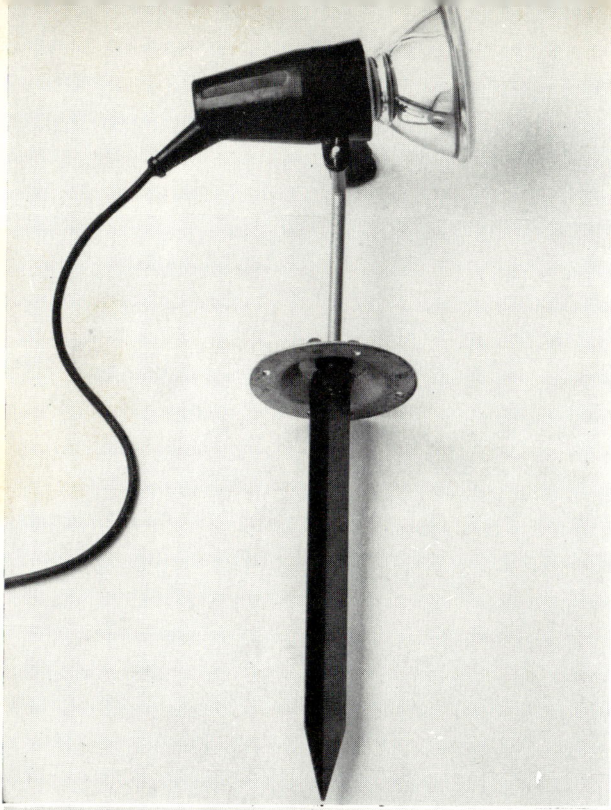

29 Weatherproof garden lighting unit with earth spike. Lamp is 100W PAR (D.R. Illuminations)

30 Low voltage spot lamp with motor operated colour change wheel (Lumitron)

31 Cylinder downlighters and reflector lamps effectively used to light a modern living room (Thorn)

spare LES lampholder but such testing is not always necessary. A feature of the lilliput bulbs is that when one fails on a series string it goes out but does not put out the whole festoon. The dud bulb is seen immediately because it is the only bulb on the festoon which is not alight.

The voltage of the bulbs depends on the number of lamps in the string. This applies whether they are MES bulbs or LES type. A 12-lamp string will usually have 20 volt bulbs; a 20-lamp string 12 volt bulbs; and in a string having a great number of bulbs the voltage rating of each bulb will be correspondingly lower – that is, a 70-lamp string uses 3.8 volt bulbs.

Replacement bulbs must be of the correct voltage for the particular string. Fit a lower voltage bulb and it will burn out; fit a higher voltage and the bulb will not light, but may reduce the light output of the other bulbs by restricting the current flow in the circuit.

When a lilliput bulb fails it should be replaced as soon as possible. When two fail, immediate replacement is essential or other bulbs will fail because voltage increases as each bulb goes. The result is the expensive failure of the entire string.

A rough check of the lamp ratings can be made by dividing the number of lampholders into 240, which is the mains voltage. For example, with a 20-lamp string; 240 volts: 20 = 12 volts per bulb. Olive-shaped bulbs and illuminated characters having MES lamp caps usually have the voltage and the watts stamped on the cap but not so the lilliput bulbs – spares being sold in packs of five with the voltage rating marked on the pack.

Some strings include a flashing bulb so the string switches on and off. Flasher bulbs can be purchased separately but they must be of the same voltage as the string. Sometimes a flasher can cause interference with television and radio reception – one's own and also one's neighbours'. An independent flasher fitted with a suppressor that sells for about .75p. can be fitted to the flex of most strings.

Always test the string before use, then disconnect from the outlet and begin to arrange it on the tree. Start at the top and work down and around, placing the bulbs first on the inner boughs close to the trunk, then moving out to the tips of the bough and so on, to the bottom of the tree. When the lights are in place, hang the other decorations on the tree.

Why restrict your Christmas lighting to a tree? Consider decorative lighting for the Christmas table. Why not a change of lamp in the ceiling fitting – a coloured lamp such as a pink pearl? Or a table decoration that uses a lamp in its centre can look most attractive. If possible, place the table against the wall so the lights may be connected to a socket without trailing wires.

Take care not to overload sockets. If other lights dim when you switch on then you may have to disconnect other electrical appliances.

APPENDIX

ADDRESSES OF SOME OF THE PRINCIPAL MANUFACTURERS

PLUGS, SWITCHES, ACCESSORIES:

Ashley Accessories Ltd., Ulverston, Lancashire.

BICC 21 Bloomsbury Street, London, W.C.1.

J. A. Crabtree & Co. Ltd., Lincoln Works, Walsall, Staffs.

MEM (Midland Electric Manufacturing Co. Ltd.), Reddings Lane, Tyseley, Birmingham.

MK Electric Ltd., Shrubbery Road, Edmonton, London, N.9.

Nettle Accessories Ltd., Warren Street, Stockport, Cheshire.

Rock Electrical Accessories Ltd., Rock Works, Commerce Road, Brentford, Middlesex.

Tenby Electrical Accessories Ltd., 19–21 Warstone Lane, Birmingham 18.

LAMP MANUFACTURERS:

Associated Lighting Ltd., Lowerhouse, West Bollington, Nr. Macclesfield, Cheshire.

British Electric Lamps Ltd., Spencer Hill Road, Wimbledon, London, S.W.19.

British Lighting Industries Ltd., Thorn House, Upper St. Martin's Lane, London, W.C.2.

Including: A.E.I. Lamp & Lighting Co. Ltd., Atlas Lighting Ltd., Ekco Lighting Ltd., Omega Lamp Works Ltd.

Ceag Ltd., Queens Road, Barnsley, Yorks.

Crompton Parkinson Ltd., 39–41 Bridge Street, Northampton.

Cryselco Ltd., Kepston Works, Bedford.

Endura Lamps Ltd., Horsforth Mill, Low Lane, Horsford, Nr. Leeds.

Nore Electric Co. Ltd., 461 Southchurch Road, Southend-on-Sea, Essex.

Osram (GEC) Ltd., P.O. Box 17, East Lane, Wembley, Middlesex.

Including: Ascot Lamps & Lighting Ltd., GEC Street Lighting Ltd.

Philips Electrical Ltd., Century House, Shaftesbury Avenue, London, W.C.2.

Including: Kingston Lamp Co. Ltd., Luxram Electric Ltd., Stella Lamp Co. Ltd.

Pope's Electric Lamp Co. Ltd., Elasta House, 16 Bromells Road, London, S.W.4.

FITTINGS

Allom Heffer & Co. Ltd., 17 Montpelier Street, Knightsbridge, London, S.W.7.

The Benjamin Electric Ltd., Station House, Darkes Lane, Potters Bar, Hertfordshire.

C. M. Churchouse Ltd., Lichfield Road, Brownhills, Walsall, Staffs.

Conelight Ltd., Longmead, Shaftesbury, Dorset.

Courtney, Pope (Electrical) Ltd., Amhurst Park Works, Tottenham, London, N.15.

Crompton Parkinson Ltd., Crompton House, 39–41 Bridge Street, Northampton.

Dernier & Hamlyn Ltd., Willowbank Wharf, Ranelagh Gardens, London, W.6.

Falks Ltd., 91 Farringdon Road, London, E.C.1.

H. & H. Electrical (London) Ltd., 253 Kilburn Lane, London, W.16.

Heyes & Co. Ltd., Water Heyes Electrical Works, P.O Box 60, Wigan, Lancs.

Holophane Ltd., P.O. Box 36, Bond Avenue, Bletchley, Bucks.

Hailwood & Ackroyd Ltd., 11 Bruton Street, London, London, W.1.

Hume Atkins & Co. Ltd., Icknield Way, Letchworth, Herts.

Linolite Ltd., 142 Liverpool Road, London, N.1.

Lumitron Ltd., 33–34 Alfred Place, London, W.C.1.

Merchant Adventurers Ltd., 231 Tottenham Court Road, London, W.1.

Multilumina Manufacturing Co. Ltd., 30 Norfolk Place, London, W.2.

Osram (GEC) Ltd., P.O. Box 17, East Lane, Wembley, Middlesex.

Philips Electrical Ltd., Century House, Shaftesbury Avenue, London, W.C.2.

Kenneth Pratt & Co. Ltd., Charterhouse Square, London, E.C.1.

Plus Lighting Ltd., 2 Patshull Road, London, N.W.5.

Rotaflex (Great Britain) Ltd., Rotaflex House, City Road, London, E.C.1.

Frederick Thomas & Co. Ltd., Everton Buildings, Stanhope Street, London, N.W.1.

Thorn Lighting Ltd., Thorn House, Upper Saint Martin's Lane, London, W.C.2.

Troughton & Young (Lighting) Ltd., 127 Sloane Street, London, S.W.1.

OUTDOOR LIGHTING

DR Illuminations, Harper Road, Wythenshaw, Manchester, 22.

Elsworthy Electronics, 21–31 Broadley Terrace, London, N.W.1.

A LIGHT SHOW

Mr. Resistor, 19 Heathmans Road, Parsons Green, London, S.W.6.

Chromosound, 88 Great Titchfield Street, London, W.1.

SOME USEFUL NAMES AND ADDRESSES

Association of Public Lighting Engineers, Buckingham Court, 78 Buckingham Gate, London, S.W.1.

British Standards Institution, 2 Park Street, London W1Y 4AA.

The Council of Industrial Design, 28 Haymarket, London, S.W.1.

Decorative Lighting Manufacturers & Distributors Association, Sardinia House, 52 Lincolns Inn Fields, London, W.C.2.

The Electricity Council, Marketing Department, Trafalgar Buildings, 1 Charing Cross, London, S.W.1.

ALSO your local Electricity Board whose address is in your local telephone directory.

Electrical Contractors Association, 55 Catherine Place, London, S.W.1.

Illuminating Engineering Society, York House, Westminster Bridge Road, London, S.E.1.

Institution of Electrical Engineers, Savoy Place, Victoria Embankment, London, W.C.2.

Lighting Industries Federation, 25 Bedford Square, London, W.C.1.

National Inspection Council for Electrical Installation Contracting, Trafalgar Buildings, 1 Charing Cross, London, S.W.1.

INDEX